High Voltage Electricity Installations

RSP SERIES IN HIGH VOLTAGE POWER TRANSMISSION

Series Editor: **Dr V. T Morgan,** *CSIRO Division of Applied Physics, Australia*

Corona Performance of High-Voltage Transmission Lines
P. Sarma Maruvada

Electromagnetic Transients in Power Systems, Second Edition
Pritindra Chowdhuri

High Voltage Electricity Installations: A Planning Perspective
Stephen A. Jay

High Voltage Electricity Installations

A Planning Perspective

Stephen A. Jay
Sheffield Hallam University, UK

John Wiley & Sons, Ltd **Research Studies Press Limited**

Published by John Wiley & Sons, Ltd., The Atrium, Southern Gate, Chichester,
West Sussex PO19 8SQ, England
Telephone (+44) 1243 779777

Email (for orders and customer service enquiries): cs-books@wiley.co.uk
Visit our Home Page on www.wiley.com

This Work is a co-publication between Research Studies Press Limited and John Wiley & Sons, Ltd.

Other Wiley Editorial Offices

John Wiley & Sons Inc., 111 River Street, Hoboken, NJ 07030, USA

Jossey-Bass, 989 Market Street, San Francisco, CA 94103-1741, USA

Wiley-VCH Verlag GmbH, Boschstr. 12, D-69469 Weinheim, Germany

John Wiley & Sons Australia Ltd, 42 McDougall Street, Milton, Queensland 4064, Australia

John Wiley & Sons (Asia) Pte Ltd, 2 Clementi Loop #02-01, Jin Xing Distripark, Singapore 129809

John Wiley & Sons Canada Ltd, 22 Worcester Road, Etobicoke, Ontario, Canada M9W 1L1

Wiley also publishes its books in a variety of electronic formats. Some content that appears in print may not be available in electronic books.

Library of Congress Cataloging-in-Publication Data

Jay, Stephen A.

High voltage electricity installations: A planning perspective / Stephen A. Jay.

p. cm.

Includes bibliographical references and index.

ISBN-13 978-0-470-03016-5 (cloth: alk. paper)

ISBN-10 0470-03016-X (cloth: alk. paper)

1. Electric apparatus and appliances. 2. High voltages. I. Title: A Planning Perspective

TK452.J39 2006

333.793'2--dc22

2006005571

British Library Cataloguing in Publication Data

A catalogue record for this book is available from the British Library

ISBN-13 978-0-470-03016-5 (HB)
ISBN-10 0470-03016-X (HB)

Typeset in 10/12pt Times New Roman by Laserwords Private Limited, Chennai, India

This book is printed on acid-free paper responsibly manufactured from sustainable forestry in which at least two trees are planted for each one used for paper production.

Table of Contents

Preface

Global warming and the increasing effects of climate change are one of the most pressing challenges of our time. The adoption of renewable energy in a clean, scalable and cost-effective manner is key to a net zero transition globally, besides playing a pivotal role in many national agendas around the world. At the same time, we have seen a revolution in artificial intelligence (AI) in recent years, powered by generative AI models that match human performance in multiple domains, driven by advances in machine learning (ML) and often Internet of Things (IoT) sensor networks. Similarly, it is expected that AI can play a crucial role in scaling and optimizing renewable energy systems, enhancing predictive capabilities, supporting decisions in Operations & Maintenance (O&M), enabling smart structural health monitoring and accelerating deployment at scale.

This book sits at the intersection of these two pressing themes: the global shift towards renewable energy and the evolving capabilities of AI. Our primary focus is on showcasing how AI research, often developed in academic settings, can be successfully translated into real-world applications in the renewable energy sector. We present case studies of successful collaborations between academia and industry, or the public sector, and focus specifically on how barriers to innovation can be overcome, enabling impactful knowledge exchange and transfer. While most case studies are centered around offshore wind and solar energy, we also incorporate perspectives on life cycle assessment, socioeconomic barriers to adoption and training needs for a growing renewables workforce.

This book was developed from the scope of a workshop on "Facilitating the Renewables Transition" held at ICLR in 2022 (see https://iclr.cc/virtual/2022/social/8733), which brought together leaders from academia, industry, and public sector organizations to identify bottlenecks and pathways for the effective integration of AI in renewable energy. One of the key takeaways from the event, which has formed the basis for much of this book, is that a considerable gap remains between cutting-edge AI research and its deployment in industrial practice. While we summarized initial results in a short contribution to Patterns, this edited volume aims to look beyond highlighting challenges and offering solutions.

A key reason for this gap is the lack of transparency associated with state-of-the-art neural network architectures. While often highly accurate in their predictions, leading deep learning models still operate as black boxes, which leads to a reluctance by stakeholders to adopt them in safety-critical or regulatory environments. A second bottleneck is the lack of high-quality datasets that are openly available for academic research, which creates a shortage of high-TRL prototypes, performance benchmarks, and agreed standards. These issues are exacerbated in newer or geographically remote renewable energy sites, where data collection is logistically difficult and often costly.

This book aims to go beyond highlighting these challenges and contributing to a roadmap to overcoming them. We present success stories of how barriers like model opacity, data scarcity, or lack of trust in academic tools were addressed through collaboration, transparency, and knowledge sharing. Each case study is designed to be transferable, offering not only insights into what was done, but how and why it worked, and how similar methods might be applied in other settings.

The chapters in this volume fall into two broad categories. First, we present detailed accounts of successful collaborations between academia and industry or third-sector stakeholders. These chapters discuss the state of research prior to collaboration, the methodologies adopted to address a shared challenge, and the resulting outcomes and impacts. We focus on bringing out the aspects that made the collaboration successful and discuss pathways to transfer.

Second, we feature forward-looking contributions that explore promising new AI methodologies. These chapters emphasize the need for AI systems that are not only high-performing but also safe, reliable, transparent, and trustworthy, which we see as core requirements for practical adoption. Here, we explore aspects like explainable AI, multimodal algorithms, cybersecurity, and other recent advances that hold promise for greater real-world adoption.

We hope that this book will serve not only as a resource but also as a catalyst, inspiring researchers, engineers, policymakers, and business leaders to come together and push the boundaries of research and technology in renewable energy, enabled by the ever-growing field of AI. By bridging the divide between research and practice, we move one step closer to a sustainable energy future—powered by both clean technologies and intelligent systems.

Nina Dethlefs and Joyjit Chatterjee

List of Contributors

Tuğçe Aker
Product Department
Faradai Limited
London
UK

James Akrigg
ARMSA Academy/Cloudy Software Limited
Manchester
UK

John Aliu
Engineering Education Transformation Institute
University of Georgia
Athens, GA
USA

Saumya Bhatnagar-Knox
School of Computing and Engineering
University of Huddersfield
Queensgate, Huddersfield
UK

James Carroll
Department of Electronic & Electrical Engineering
University of Strathclyde
Glasgow
UK

Sandra Castellanos-Paez
School of Computing and Engineering
University of Huddersfield
Queensgate, Huddersfield
UK

Manu Centeno-Telleria
Fluid Mechanics Department
Mondragon University
Arrasate
Spain

Joyjit Chatterjee
Department of Computer Science
Loughborough University
Loughborough, Leicestershire
UK

EPAM Systems
London
UK

Elizabeth J. Cross
Dynamics Research Group
School of Mechanical
Aerospace and Civil Engineering
University of Sheffield
Sheffield
UK

Lekan Damilola Ojo
Department of Building and Real Estate
The Hong Kong Polytechnic University
Hong Kong
China

Nina Dethlefs
Department of Computer Science
Loughborough University
Loughborough, Leicestershire
UK

Language and Data Research Group
Loughborough University
Loughborough, Leicestershire
UK

Anubhav Gautam
Amity School of Engineering
and Technology
Amity University
Noida, Uttar Pradesh
India

Deepak Gautam
School of Management and
Entrepreneurship
Shiv Nadar Institution of Eminence
Noida, Uttar Pradesh
India

Rongge Guo
School of Traffic and Transportation
Beijing Jiaotong University
Beijing
China

Hanu Priya Indiran
Department of Engineering
University of Cambridge
Cambridge
UK

Kunal Jindal
Amity School of Engineering
and Technology
Amity University
Noida, Uttar Pradesh
India

Matthew R. Jones
Dynamics Research Group
School of Mechanical
Aerospace and Civil Engineering
University of Sheffield
Sheffield
UK

Mingxin Li
Department of Electronic & Electrical
Engineering
University of Strathclyde
Glasgow
UK

Department of Civil Engineering
The University of Tokyo
Tokyo
Japan

Weijiang Lin
Dynamics Research Group
School of Mechanical
Aerospace and Civil Engineering

University of Sheffield
Sheffield
UK

Jonathan LeSage
Application Engineering Group
MathWorks Inc.
Plano, TX
USA

Rakesh Maharaj
ARMSA Academy
Manchester
UK

Anu Mehra
Amity School of Engineering and Technology
Amity University
Noida, Uttar Pradesh
India

Robin S. Mills
Dynamics Research Group
School of Mechanical
Aerospace and Civil Engineering
University of Sheffield
Sheffield
UK

Dorcas T. Moyanga
Department of Quantity Surveying
Federal University of Technology
Akure
Akure
Nigeria

Oluwadamilare Olamide Ilesanmi
Department of Civil Engineering
Lakehead University
Thunder Bay
Canada

Ayodeji Emmanuel Oke
Department of Quantity Surveying
Federal University of Technology
Akure
Akure
Nigeria

Francesco Percassi
School of Computing and Engineering
University of Huddersfield
Queensgate, Huddersfield
UK

Daniel J. Pitchforth
Dynamics Research Group
School of Mechanical
Aerospace and Civil Engineering
University of Sheffield
Sheffield
UK

Keith Worden
Dynamics Research Group
School of Mechanical
Aerospace and Civil Engineering
University of Sheffield
Sheffield
UK

Oladotun Nathanael Agboola
Department of Construction and Real Estate Management
Metropolia University of Applied Sciences
Helsinki
Finland

Megan Robinson
School of Computing and Engineering
University of Huddersfield
Queensgate, Huddersfield
UK

Callum Rothon
Language and Data Research Group
Loughborough University
Loughborough, Leicestershire
UK

Aura Centre for Doctoral Training
Energy & Environment Institute
University of Hull
Hull, East Riding of Yorkshire
UK

Garima Sharma
Application Engineering Group
MathWorks Inc.
Plano, TX
USA

Mauro Vallati
School of Computing and Engineering
University of Huddersfield
Queensgate, Huddersfield
UK

1

AI for Renewables: Addressing Operational, Engineering, and Socioeconomic Adoption Challenges

Joyjit Chatterjee[1,2] and Nina Dethlefs[2]*

[1] *EPAM Systems, London, UK*
[2] *Department of Computer Science, Loughborough University, Loughborough, Leicestershire, UK*

*Corresponding author: Joyjit Chatterjee; joyjit_chatterjee@epam.com

1.1 Introduction

With the growing global push toward decarbonization and aspiration toward net-zero targets, there has been a rapid rise in the adoption of renewable energy sources such as solar, wind, hydropower, etc. Despite their immense promise as non-conventional sources of energy, the inherent nature of such resources introduces significant variability and complexity into the domain of energy systems, which calls for systematic planning and Operations and Maintenance (O&M) strategies. There has been tremendous growth in the renewables industry in recent times, with the renewable energy market estimated to jump from USD 1.34 trillion in 2024 to an almost 4*x* market value of 5.62 trillion by 2033 (Shuliak, 2025), with AI being an integral force driving this scale-up.

Over the last few years, AI has transitioned from just being a buzzword in the renewables community to playing a game-changing role in data-driven decision support by helping optimize generation, managing energy demand, and enhancing reliability of such intermittent sources of energy. This has largely been driven by the Industry 4.0 era, including advances in Internet of Things (IoT) and high-quality sensors (e.g. from Supervisory Control & Data Acquisition – SCADA system) that can capture high-frequency temporal data (Qays et al., 2022; Ziane et al., 2022). The multitude of recent advances in AI, in particular Generative AI (GenAI) and Large Language Models (LLMs), have shown promise for tackling sector-specific challenges. This has been exhibited through multiple research publications in academia as well as proof of concepts in industry. The vast and

Renewable Energy Transition with Artificial Intelligence: Challenge-driven Solutions, First Edition. Nina Dethlefs and Joyjit Chatterjee. Published 2026 by John Wiley & Sons, Inc.

complex nature of this domain is highlighted in Figure 1.1 – ranging from application of (i) algorithm-centered AI techniques like XGBoost, CatBoost, LSTM, reinforcement learning, meta-heuristic techniques such as gray-wolf optimization, etc., (ii) an operational cluster of aspects like microgrids, MPPT, HVDC, predictive maintenance, and several system-level technical terms, and (iii) socioeconomic connections to energy trading, sustainability, carbon emissions, GDP, etc. This showcases that the overall landscape is not just growing in the application of computational techniques, but also already has a broad focus on system engineering and financial as well as operational impacts.

However, the deployment of AI in real-world scenarios in production comes with several challenges, which range from the need for trust and explainability in model decisions to optimal performance across widely varying regions and conditions. By selectively analyzing Web of Science for 7200 conference and journal articles between 2021 and 2025 that specifically focus on AI in the renewables domain, we can see a rapid rise in their volume (CAGR 42% up to 2024 – note that indexing for the current year 2025 is not yet complete at the time of writing this chapter). Figure 1.2a relates to the number of papers involving AI and OSW since 2021. Figure 1.2b shows the top 10 venues of publication, clearly highlighting high-impact energy/AI-focused outlets. Figure 1.2c enunciates the most frequent WoS subject categories, affirming the continued focus on renewables + AI.

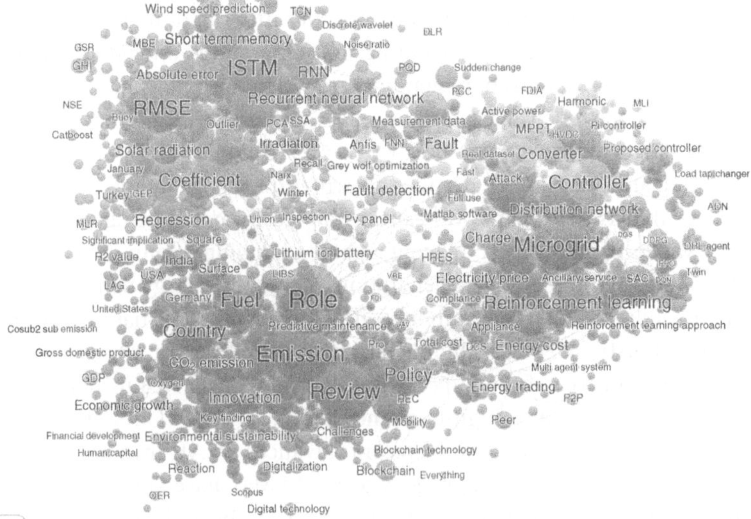

Figure 1.1 Network visualization in VOSviewer for term co-occurrence (2021–2025).

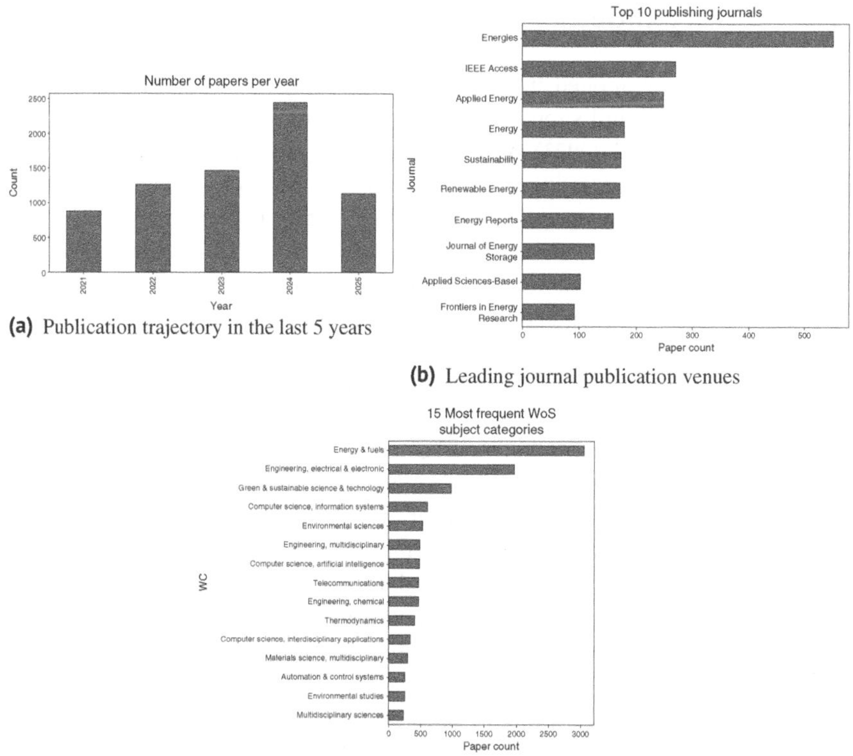

(a) Publication trajectory in the last 5 years

(b) Leading journal publication venues

(c) Most prevalent subject areas in WoS

Figure 1.2 Scientometric insights excluding network visualization.

This chapter provides an introduction to the various ways AI is contributing to the renewable energy transition through challenge-driven solutions. We discuss prominent use-cases and challenges uncovered through foundational research across industry and academia over the years, highlighting explainable AI (XAI) and transferable models, GenAI for efficient energy planning, and methods that facilitate grounding AI models with domain-specific knowledge across energy sources. The chapter also discusses emerging applications – like AI-augmented smart grids, predictive maintenance, energy forecasting, and climate-resilient infrastructure planning, whilst also identifying critical areas that warrant further R&D. We aim to address with this accessible yet technically grounded overview the current gap that hinders adoption of AI more widely in the renewables sector, by bringing together subject-matter experts, engineers and technicians in renewables, AI specialists, researchers, and industry leaders on the same page regarding the present and future direction of AI for facilitating a smoother transition to renewable energy.

1.2 Opportunities and Challenges

AI is being widely applied in the renewables domain for processing massive streams of big data and supporting timely decisions in complex, safety-critical systems (Alsaigh et al., 2023). Machine Learning (ML) models play an integral role these days in forecasting renewable power generation, predictive maintenance, optimizing grid operations and stability, and facilitating control of distributed resources (Mousavi et al., 2025; Carroll et al., 2016; Chatterjee and Dethlefs, 2021a). For instance, in the wind industry, deep learning models, e.g. Long Short-Term Memory (LSTM), networks have become a de facto standard for short-term forecasting of wind power, exhibiting their unique capability to capture nonlinear weather-power associations significantly better than traditional statistical techniques (Yassen et al., 2025). The industry has also witnessed some notable breakthroughs that have been in the limelight – such as a collaboration between DeepMind and the UK's National Grid in using AI to optimize wind energy production that led to 20% higher output than traditional methods; also IBM Research's demonstration of AI for improved solar power forecasting, more efficient grid management, and end-to-end integration, etc. (Ukoba et al., 2024).

There is growing interest in utilities and grid operators for leveraging AI to improve smart grid management – including balancing supply and demand in real-time, more efficient integration of distributed energy resources, and better accuracy in fault detection and response to outages. A plethora of techniques have been leveraged already in the industry to tackle the intermittent nature of renewable energy systems, such as Gaussian Process Regression, Support Vector Regression, Artificial Neural Networks, etc. (Ukoba et al., 2024). There are a plethora of benefits that AI brings to the table: reduced downtimes and enhanced operational efficiency, lower costs in O&M, and higher reliability for renewable energy systems. According to domain experts, AI has immense potential in reducing the costs and accelerating the processes associated with designing, licensing, deploying, operating, and maintaining energy infrastructure, with savings that could span hundreds of billions of dollars (Kooei, 2024).

While such opportunities are undoubtedly evident, there are several key challenges that come as barriers that prevent AI from realizing its transformative potential in the energy sector. One of the most critical issues is the *black-box* nature of most modern AI models, particularly deep learners. While neural nets often achieve high accuracy owing to the inherently complex and nonlinear nature of the data in this domain, the multiple variables and the dynamic interactions between them make accurate and reliable modeling challenging. In the energy sector, stakeholders (operators, engineers/technicians, and regulators) are often reluctant to adopt such complex models, which cannot offer clear explanations behind their decisions (Alsaigh et al., 2023). Another key challenge is the lack of transferability – ML models trained with knowledge from a specific context often struggle to generalize to other new, unseen contexts that can arise due to differences in geography, varying climatic conditions, unique grid configurations, etc. This raises the pivotal need for addressing model transferability, scalability, and the ability to generalize across varying system configurations and regions.

Lack of sufficient data for training ML models is another practical challenge, e.g. for new solar or wind farm developments that have little to no historical data in place to train reliable models. Additionally, incomplete or inaccurate data (e.g. anomalous values from SCADA sensors in a turbine, missing power generation metrics in a solar panel, etc.) can produce models that learn sub-optimally and predict with limited accuracy, which undermines their utility in real-world deployment. Whilst strategies like transfer learning (TL) and domain adaptation have become popular in the last few years to tackle this concern (Cheng et al., 2025), they come with their own set of complexities that hinder the ability to ensure that adapted models remain accurate and unbiased in new scenarios. Additionally, the latest advances in GenAI and LLMs, while promising, come with an inevitable need for grounding the foundation models in domain-specific knowledge without which the models cannot be trusted to generate valid and useful outputs in the renewables industry. Without any physical constraints or system-specific expertise, these powerful models may provide fluent and plausible-sounding recommendations that are actually technically flawed and inaccurate. In the next section, we highlight some of the high-priority areas that directly address the constraints highlighted thus far.

1.3 Current High-priority Areas

1.3.1 Explainability and Trust in AI for Renewables

As AI models take on more autonomous roles in facilitating decision support in the renewables industry – from controlling grid assets to forecasting power produced, stakeholders require a clear understanding of why a model makes certain predictions or recommendations, particularly in high-stakes, safety-critical environments like offshore wind farms and electricity grids. Without a mechanism for transparency in place, stakeholders will not be able to trust model advice, thus hampering adoption and leading to significantly costly outcomes, e.g. overreacting to false alarms or missed detection of valid anomalies. For instance, if a model recommends curtailment of wind power generation without providing a clear rationale, the engineers would be skeptical to act on such alerts owing to the risk of misjudgment (Paul et al., 2024). Developing interpretable models thus requires them to be in line with human operators' domain knowledge as well as regulatory requirements in the energy sector.

Researchers are increasingly augmenting traditionally black-box models, especially deep learners with XAI techniques to foster trust and confidence amongst humans by making the models' learning process and predictions more understandable (Shadi et al., 2025). A wide spectrum of XAI techniques has been explored in recent times in the renewables domain – ranging from basic feature importances (e.g. in decision tree models like XGBoost) to more sophisticated visualization of reasoning in neural nets (e.g. via attention mechanism in models like Transformers). Local post-hoc explanation techniques like SHapley Additive exPlanations (SHAP) and Local Interpretable Model-Agnostic Explanations (LIME) have become particularly popular for energy forecasting models, since they can attribute model predictions to specific input features corresponding to any particular forecast. For instance, SHAP has recently been applied to deep learning models for wind power forecasting, attributing the key environmental factors (such as wind speed, temperature, etc.) that influence the predictions (Yassen et al., 2025). Such studies have found that XAI techniques can provide actionable and human-interpretable insights to stakeholders in the renewables domain, enhancing the wider acceptance and adoption of complex AI models.

Another interesting area with growing interest is in leveraging intrinsically interpretable ML models. For instance, physics-guided neural nets (e.g. using power flow equations) and rule-based ML models (e.g. with if-else conditional rules) can be embedded with domain-specific knowledge, which can help them to justify decisions in terms more familiar to subject-matter experts (SMEs). In relationship to smart grids, researchers have mentioned that ML models should ideally be *explainable and interpretable* by design, to adhere to complex regulatory and safety requisites. An explainable decision support system should be able to trace how it arrived at a recommendation, e.g. by explaining that a predicted drop in solar farm power output is due to incoming overhead clouds observed via satellite data, supplemented with the model's learnt patterns for similar weather conditions that may have occurred historically. Such clarity is integral to ensuring accountability and providing a mechanism for human operators to validate the model's outputs.

Beside academia, there have been focused government and industry initiatives that heavily emphasize the need for explainability in AI models within the energy sector. The National Institute of Standards and Technology (NIST) in the United States has specifically defined certain key principles for trustworthy AI that include validity and reliability; safety, security and privacy; fairness and bias management; accountability and transparency, explainability and interpretability (Choi et al., 2024). National laboratories are now developing AI systems that enable grid control and are inherently embedded with explainability provisions. The eGridGPT decision support model from the National Renewable Energy Laboratory (NREL) is a notable example, as it incorporates a combination of physics-informed modeling and XAI to ensure that the rationale behind the model's recommendations are clearly understandable by grid operators. By providing the ability to compare AI predictions against physical simulations for power flow and providing reason codes for different anomalies (e.g. violations in voltage limit), eGridGPT aims to foster trust and confidence amongst grid operators.

Explainability and trust are a clear challenge-driven priority in the renewables industry. While notable progress has been made to enunciate model decisions, there is more concrete continued research required to tailor XAI methods that are tailored to the complex nature of engineering systems like wind turbines and the development of industry-wide standards for AI transparency. With predictions that are actually understandable at the ground level, human-AI collaboration will play a key role in unlocking wider adoption of such models in improving O&M.

1.3.2 Model Transferability and Generalization

The development and deployment of renewable energy systems globally varies widely, since different sources of energy are adapted across topographies, grid infrastructures, climatic conditions and socioeconomic backgrounds. An ML model that is trained on one kind of dataset (e.g. from an offshore wind farm in Europe) may exhibit sub-optimal performance when used for inference in a different context (e.g. an onshore wind farm in North America), owing to drift in the operating conditions and the associated dataset probabilistic distributions. Stakeholders in the industry require AI solutions to not just be *one-off* that only work in a narrow set of constraints, but can also be scaled and transferred to new contexts. The goal in achieving such generalization capabilities has become a significant area of focus in recent times.

A popular approach for accomplishing this is Transfer Learning (TL), an area of ML that aims to adapt models trained with data from a source domain into a target domain, such that the pre-trained model can serve as the foundation for the new domain. While traditional ML often assumes that training and test data would come from a similar feature space, this assumption often does not hold true in real-world energy systems. TL facilitates the ability to transfer the knowledge learnt from a certain task or region into a different one, thus relaxing the requirements for the quantity of new data needed for decision support in the new domain. There has been growing interest in TL amongst researchers to address cross-domain generalization, largely driven by the common challenge of improving the accuracy of predictive models for fault diagnosis and power forecasting in situations with scarce data (Al-Hajj et al., 2023).

TL has been explored in multiple applications, but most predominantly in wind and solar power forecasting tasks as well as areas like wind resource assessment, since this enables more efficient planning and resource allocation

(Ying et al., 2023). Any variances in critical parameters (e.g. wind speed, terrain effects, etc.) between the source and target domains are adjusted for via domain adaptation techniques. The hidden layers of a deep neural net can be shared between the source and target domain (e.g. for two wind farms in different geographical locations) to achieve better performance than feasible with traditionally independent training (Chatterjee and Dethlefs, 2020). More recent approaches for TL combine multi-task learning with fine-tuning to achieve improved accuracy. For instance, an ML model could be trained with historical data from multiple existing solar farms, and then fine-tuned with limited data from a new solar farm. Another popular approach is to leverage synthetic data – a base model is initially fine-tuned on a small dataset with real-world context from the new site, which is then used for generating additional training data samples (e.g. via GenAI models, atmospheric models for weather simulations, etc.). A recent study in the wind industry applied such a technique toward domain-invariant icing prediction in wind turbine rotor blades, by leveraging synthetic data augmentation with GenAI to improve generalization (Chatterjee et al., 2023). The authors utilized neural style transfer to augment rotor blade RGB images with synthetic ice patterns, so that the icing prediction model could learn transferable features in the images across different wind farms as well as new blade designs and weather conditions (e.g. summer vs winter). They demonstrated that the model could perform icing detection with high accuracy even on unseen wind farms, thus supporting scalable O&M across different contexts. Another notable study (Tang et al., 2024) has leveraged a combination of adversarial training and temporal convolution to achieve optimal wind power forecasting in new domains by ensuring a distribution-centric perspective. Novel techniques such as hybridizing stacking and TL for ultra-short-term wind power forecasting have also been showcased (Cheng et al., 2025), wherein, dimensionality reduction is initially performed via Principal Component Analysis (PCA), and then several base learners (e.g. LSTM, BiLSTM, GRU, etc.) and attention-based models (e.g. LSTM with attention) are combined with a stacking ensemble approach.

While these use cases are promising, there remain challenges in ensuring that the AI is truly robust and scalable for all types of real-world conditions, not just a narrow set of context types. Common situations may include, e.g. deployment of solar panels at a new site that has no historical power generation data to enable time-series forecasting; developing predictive models for a wind farm that has not been in operation for long and only has few months of condition-based monitoring (CBM) data, etc. To tackle instances with no historical data, physics-based or GenAI simulations or leveraging data from similar sites (e.g. in terms of topography and environmental conditions) could be used to train an ML model – however, it is very challenging to ensure similarity in all respects holistically and there is a possibility of introducing inadvertent bias from one context into another.

When dealing with small historical datasets for training models, the data may fail to capture nuances and seasonal variability, which can lead a TL model to make poor decisions (e.g. during unseen seasons and weather conditions).

1.3.3 Grounding AI Models to Domain-specific Operational and Engineering Knowledge

With the advent of LLMs, there has been a rapidly rising interest in the renewables industry, particular amongst senior stakeholders (e.g. C-suite executives) in applying GenAI across the board. Much of this interest is driven by the current hype in GenAI across other industries, ranging from healthcare and life sciences, to finance and retail (Xi, 2024). While foundation models (e.g. GPT-o3, Claude 4 Opus, etc.) are powerful and adept at general-purpose tasks (e.g. searching information from the web and providing insights, summarizing documents, pursuing deep research on specific topics, etc.), they do not have business-specific context and domain knowledge grounded in them since much of this is not publicly available (e.g. O&M manuals, work orders, historical SCADA data and alarm logs, etc.). Using such models directly in safety-critical contexts like the energy sector entails massive risks of hallucinations and non-factual information, relying on which can lead to more damage than benefits for operators.

Knowledge graphs (KGs) are a structured network of entities (nodes) and their associated relationships, which in the context of renewables can be real-world system components, e.g. wind turbine sub-components, solar panel fault types, alarms in a hydro-power plant, etc. Nodes can also contain rich properties (e.g. detailed descriptions of components, thresholds for anomalies, images, etc.).

Past research has demonstrated that integrating structured knowledge into AI models with KGs helps enable accurate, explainable and context-aware decision-making (Ontoforce, 2024). Some concrete ways in which the renewables industry can leverage KGs include (but are not limited to):-

- Modeling relationships between environmental conditions and energy generation
- Grid topology network and inter-connections in the infrastructure
- System-specific equipment details (e.g. wind turbine sub-components like gearbox, generator, etc.) and associated O&M actions (e.g. preventive actions toward fixing faults attributed to rising gearbox oil temperature) (Chatterjee and Dethlefs, 2021b)
- Regulatory constraints, health and safety requirements, and market dynamics

KGs enable reasoning, interactive querying, and integration of a variety of heterogeneous data sources but are unfortunately still at a nascent stage in the renewables domain. The key reason for this includes lack of unified standards and open-data (e.g. SCADA, alarms, maintenance logs, etc.) that hinder the development of open-source conceptual models and ontologies in this sector. Another constraint till date has been a lack of practical demonstration of use cases for O&M or decision support, and the challenge for engineers and technicians to access the domain-specific KGs since historically, querying such information has often required graph query languages (e.g. Cypher).

With LLMs now capable of writing code in such graph query languages themselves based on a user's natural language prompt, the landscape has huge potential and is starting to transition into interactive question-answering over KGs (Chatterjee and Dethlefs, 2022). Figure 1.3 depicts an example of how LLMs can be leveraged for interactive decision support via natural language conversations by O&M engineers and technicians in the wind industry – same principles can be applied to other energy systems like solar. Additionally, with advances like Model Context Protocol (MCP) (Hou et al., 2025) in limelight, it is now possible to integrate any domain-specific KG with an LLM model easily (e.g. by initiating a MCP server from the Claude foundational LLM to Neo4j Graph Database Management System) (Chatterjee and Dethlefs, 2021b).

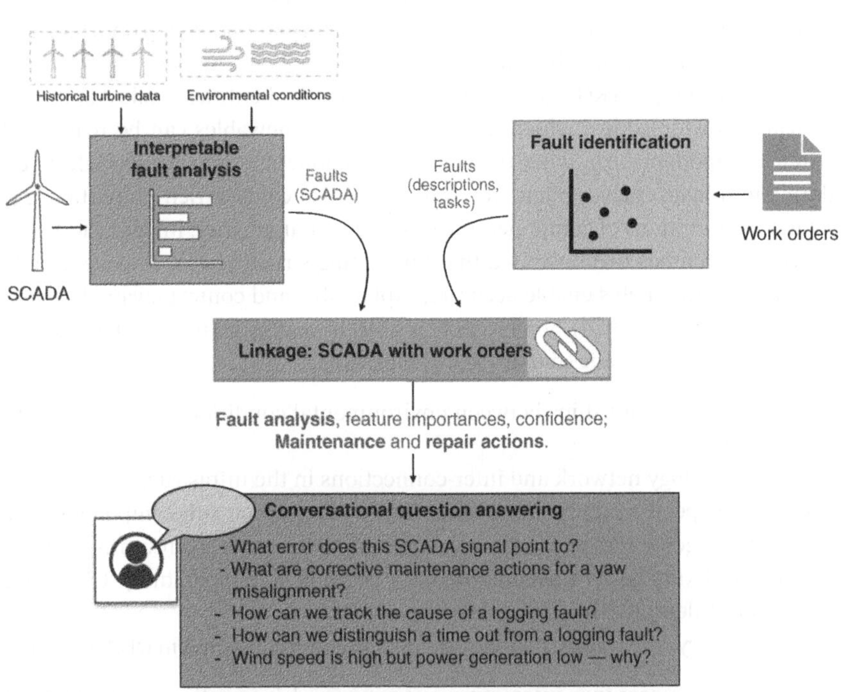

Figure 1.3 Leveraging LLMs toward conversational question-answering in the wind industry. *Source*: Adapted from Chatterjee and Dethlefs (2021b).

1.4 Nascent Areas in the AI and Renewables Domain

This chapter has articulated several ways in which academia and industry have already started to apply and benefit from the power of data science and AI, particularly in the solar and wind energy domains. However, there are still a plethora of aspects which are at an embryonic stage:-

1. Multi-agentic AI frameworks (e.g. by integrating a weather forecasting agent, financial impact modeling agent, predictive maintenance agent, offshore vessel transfer agent, etc.) that transition from current stand-alone predictive ML models for disparate areas (e.g. anomaly prediction) to a holistic human-in-the-loop decision support system that autonomously plans, reasons, and calls tools/Application Program Interfaces (APIs) to address root causes to prevent incipient issues.
2. Edge computing to enable real-time decision support and control operations with low latency, high security, and low bandwidth.
3. Reinforcement learning to support long-term energy forecasting and O&M planning, dynamic control and optimization strategies that facilitate autonomous decision-making in changing conditions (e.g. harsh weather, critical faults, etc.).
4. Blockchain integration to enable decentralized grid management.
5. Industry-wide evolution of standardized regulatory and data-centric policies and O&M frameworks, e.g. in data collection and structuring conventions from SCADA sensors, open-source KGs for offshore and onshore wind farms, etc.

1.5 Conclusion

In this chapter, we have provided a comprehensive perspective on the past, present, and future of AI in the renewables sector, particularly in the solar and wind domains, which are currently the leading clean energy sources globally. Despite the growing popularity of AI, we note that the renewables industry is currently behind other industries like healthcare and finance in adopting state-of-the-art techniques like agentic AI, grounding LLMs by integrating with KGs, transfer learning to enable scalable ML that addresses climatic variations and data heterogeneity. To accomplish this, there needs to be greater focus on large-scale sharing of open data by the industry, which will enable stronger academia–industry collaboration. Additionally, researchers need to pivot their future focus from purely accuracy to explainability and transparency, which would mean that using complex deep learning models and frameworks that leverage LLMs should only be warranted when there is sufficient operational and engineering knowledge instilled in the models. We envision that this will help translate novel innovations into real-world production-ready deployments in the energy sector.

Bibliography

Rami Al-Hajj, Ali Assi, Bilel Neji, Raymond Ghandour, and Zaher Al Barakeh. Transfer learning for renewable energy systems: A survey. *Sustainability*, 15(11):9131, 2023. https://doi.org/10.3390/su15119131.

Roba Alsaigh, Rashid Mehmood, and Iyad Katib. AI explainability and governance in smart energy systems: A review. *Frontiers in Energy Research*, 11:1071291, 2023. https://doi.org/10.3389/fenrg.2023.1071291.

James Carroll, Alasdair McDonald, and David McMillan. Failure rate, repair time and unscheduled O&M cost analysis of offshore wind turbines. *Wind Energy*, 19(6):1107–1119, 2016. https://doi.org/10.1002/we.1887.

Joyjit Chatterjee and Nina Dethlefs. Deep learning with knowledge transfer for explainable anomaly prediction in wind turbines. *Wind Energy*, 23(8):1693–1710, 2020. https://doi.org/10.1002/we.2510.

Joyjit Chatterjee and Nina Dethlefs. Scientometric review of artificial intelligence for operations & maintenance of wind turbines: The past, present and future. *Renewable and Sustainable Energy Reviews*, 144:111051, 2021a. https://doi.org/10.1016/j.rser.2021.111051.

Joyjit Chatterjee and Nina Dethlefs. Xai4wind: A multimodal knowledge graph database for explainable decision support in operations & maintenance of wind turbines. *arXiv preprint arXiv:2012.10489*, 2021b. https://arxiv.org/abs/2012.10489.

Joyjit Chatterjee and Nina Dethlefs. Automated question-answering for interactive decision support in operations & maintenance of wind turbines. *IEEE Access*, 10:84710–84737, 2022. https://doi.org/10.1109/ACCESS.2022.3197167.

Joyjit Chatterjee, Maria T Alvela Nieto, Hannes Gelbhardt, Nina Dethlefs, Jan-Hendrik Ohlendorf, Andreas Greulich, and Klaus-Dieter Thoben. Domain-invariant icing detection on wind turbine rotor blades with generative artificial intelligence for deep transfer learning. *Environmental Data Science*, 2:e12, 2023. https://doi.org/10.1017/eds.2023.9.

Xu Cheng, Yu Cao, Zhiyuan Song, and Chenguang Zhang. Wind power prediction using stacking and transfer learning. *Scientific Reports*, 15(1):11566, 2025. https://doi.org/10.1038/s41598-025-96262-6.

Seong Lok Choi, Rishabh Jain, Patrick Emami, Karin Wadsack, Fei Ding, Hongfei Sun, Kenny Gruchalla, Junho Hong, Hongming Zhang, Xiangqi Zhu, et al. egridgpt: Trustworthy AI in the control room. *Technical Report*, National Renewable Energy Laboratory (NREL), Golden, CO (United States), 2024. https://www.osti.gov/biblio/2352232.

Xinyi Hou, Yanjie Zhao, Shenao Wang, and Haoyu Wang. Model context protocol (mcp): Landscape, security threats, and future research directions. *arXiv preprint arXiv:2503.23278*, 2025. https://arxiv.org/abs/2503.23278.

Michael Kooei. How artificial intelligence can transform U.S. energy infrastructure, Argonne National Laboratory, 2024. https://www.anl.gov/article/how-artificial-intelligence-can-transform-us-energy-infrastructure.

Rashin Mousavi, Arash Mousavi, Yashar Mousavi, Mahsa Tavasoli, Aliasghar Arab, Ibrahim Beklan Kucukdemiral, Alireza Alfi, and Afef Fekih. Revolutionizing solar energy resources: The central role of generative AI in elevating system sustainability and efficiency. *Applied Energy*, 382:125296, 2025. https://doi.org/10.1016/j.apenergy.2025.125296.

Ontoforce. Knowledge graphs on the rise: Gartner's 2024 AI hype cycle shows their growing impact, 2024. https://www.ontoforce.com/blog/knowledge-graphs-on-the-rise-gartners-2024-ai-hype-cycle-shows-their-growing-impact.

Shuva Paul, Sanjana Vijayshankar, and Richard Macwan. Demystifying cyberattacks: Potential for securing energy systems with explainable AI. In *2024 International Conference on Computing, Networking and Communications (ICNC)*, pp. 430–434. IEEE, 2024. https://doi.org/10.1109/ICNC59896.2024.10556212.

Md Ohirul Qays, Musse M. Ahmed, M. A. Parvez Mahmud, Ahmed Abu-Siada, S. M. Muyeen, Md Liton Hossain, Farhana Yasmin, and Md Momtazur Rahman. Monitoring of renewable energy systems by IoT-aided scada system. *Energy Science & Engineering*, 10(6):1874–1885, 2022. https://doi.org/10.1002/ese3.1130.

Mohammad Reza Shadi, Hamid Mirshekali, and Hamid Reza Shaker. Explainable artificial intelligence for energy systems maintenance: A review on concepts, current techniques, challenges, and prospects. *Renewable and Sustainable Energy Reviews*, 216:115668, 2025. https://doi.org/10.1016/j.rser.2025.115668.

Maryna Shuliak. AI in renewable energy: Use cases, benefits & solutions for 2025, 2025. https://acropolium.com/blog/artificial-intelligence-and-renewable-energy-a-guide-to-tech-sustainability/.

Yugui Tang, Kuo Yang, Yichu Zheng, Li Ma, Shujing Zhang, and Zhen Zhang. Wind power forecasting: A transfer learning approach incorporating temporal convolution and adversarial training. *Renewable Energy*, 224:120200, 2024. https://doi.org/10.1016/j.renene.2024.120200.

Kingsley Ukoba, Kehinde O. Olatunji, Eyitayo Adeoye, Tien-Chien Jen, and Daniel M. Madyira. Optimizing renewable energy systems through artificial intelligence: Review and future prospects. *Energy & Environment*, 35(7):3833–3879, 2024. https://doi.org/10.1177/0958305X241256293.

Ruichen Xi. Applications of knowledge graph in medical and financial fields: Data integration and intelligent decision-making from an interdisciplinary perspective. *Applied and Computational Engineering*, 67:300–306, 2024. https://doi.org/10.54254/2755-2721/67/20240471.

Mona Ahmed Yassen, El-Sayed M. El-Kenawy, Mohamed Gamal Abdel-Fattah, Islam Ismail, and Hossam El-Deen Mostafa. Explainable artificial intelligence for wind power forecasting model based on long short-term memory. *Neural Computing and Applications*, 37:14589–14611, 2025. https://doi.org/10.1007/s00521-025-11230-5.

Changtian Ying, Weiqing Wang, Jiong Yu, Qi Li, Donghua Yu, and Jianhua Liu. Deep learning for renewable energy forecasting: A taxonomy, and systematic literature review. *Journal of Cleaner Production*, 384:135414, 2023. https://doi.org/10.1016/j.jclepro.2022.135414.

Abderrezzaq Ziane, Rachid Dabou, Ammar Necaibia, Abdelkrim Rouabhia, Kada Bouchouicha, Nordine Sahouane, Salah Lachtar, Ahmed Bouraiou, and Ahmed Amine Larbi. An IoT platform for online monitoring of renewable energy systems. In *2022 3rd International Conference on Embedded & Distributed Systems (EDiS)*, pp. 55–60. IEEE, 2022. https://doi.org/10.1109/EDiS57230.2022.9996490.

2

Techno-economic Analysis for Offshore Renewable Energy Technologies Incorporating a Holistic O&M Model

Mingxin Li,[1,2] Manu Centeno-Telleria,[3] and James Carroll[1]*

[1] *Department of Electronic & Electrical Engineering, University of Strathclyde, Glasgow, UK*
[2] *Department of Civil Engineering, The University of Tokyo, Tokyo, Japan*
[3] *Fluid Mechanics Department, Mondragon University, Arrasate, Spain*

*Corresponding author: Manu Centeno-Telleria; mcentenot@mondragon.edu

2.1 Challenge

The global energy transition demands a substantial deployment of offshore renewable energy systems, encompassing both mature technologies, like bottom-fixed offshore wind, and less mature technologies, such as floating offshore wind (FOW), wave energy, and tidal energy. Nevertheless, the levelized cost of energy (LCoE) of FOW, wave energy, and tidal energy must be reduced for commercializing the technology. As a comprehensive approach for evaluating the performance of an industrial project, techno-economic analysis combines various essential elements, including process design, process modeling, equipment sizing, capital cost estimation, and operating cost estimation. In offshore projects and other renewable energy initiatives aligned with net-zero strategies, techno-economic analysis provides a holistic framework to assess technological feasibility, economic viability, and environmental impact. By bridging technical and economic considerations, techno-economic models incorporating a life-cycle perspective offer strategic insights to identify cost-reduction opportunities, guide sustainable energy transitions, and support the achievement of carbon neutrality goals.

The LCoE serves as a widely recognized metric for assessing and comparing the cost efficiency of different energy generation technologies (Li et al., 2024a). Accurate estimation of the LCoE for offshore renewable energy farms is crucial, as it is inherently influenced by site-specific factors, with energy yield, capital expenditures (CapEx), and operational expenditures (OpEx) being location-dependent. For instance, operating in offshore deep-sea locations with stronger and more stable winds allows for the use of larger floating wind turbines, which can potentially increase energy output. However, the CapEx of larger turbines and the OpEx in deep-sea locations also increase accordingly. CapEx estimations for offshore renewable projects are typically more robust due to the availability of established data and methodologies. In contrast, OpEx for offshore renewable energy remains a significant unknown, particularly for floating offshore wind (FOW), wave energy, and tidal energy, given the limited operational experience with these technologies. This uncertainty underscores the need for holistic operations and maintenance (O&M) characterization within techno-economic models to better inform cost estimates and support the commercial viability of these emerging energy solutions.

Holistic O&M consideration in techno-economic models involves a comprehensive analysis of the various factors that affect OpEx and energy production throughout the life cycle of offshore renewable energy projects. This includes assessing attributes such as reliability, maintainability, accessibility, and availability (Centeno-Telleria et al., 2022). Furthermore, key O&M factors, including failure rates, repair times, operational limits of vessels, maintenance strategies, and associated costs, should also be integrated into the models. By capturing these elements, techno-economic models can offer more precise insights into the operational challenges and costs, enabling stakeholders to make informed decisions that enhance the economic viability and efficiency of offshore renewable energy systems.

In this context, performing broad spatial assessments of LCoE for less mature offshore renewable energy technologies is essential for identifying the most suitable locations for deployment. These assessments allow for the comparison of LCoE across different geographical areas, taking into account local metocean conditions and infrastructure capabilities, such as ports. By integrating spatial data into the analysis, stakeholders can better understand the economic potential of various sites, leading to more strategic planning and decisions affecting the final cost.

Therefore, a key challenge for offshore renewable energy technologies lies in developing a life-cycle techno-economic model that includes holistic O&M considerations and enables broad spatial assessments. This chapter aims to address this challenge by creating a comprehensive techno-economic assessment that incorporates various O&M factors. These factors are examined in relation to system characteristics, such as reliability, maintainability, accessibility, and availability, and are subsequently integrated into the techno-economic assessment and LCoE mapping. The insights gained from the proposed techno-economic model contribute to the development of more reliable Artificial Intelligence (AI) models for predictive maintenance and decision support. The research enables the quantification of the potential returns on investment in offshore projects driven by future AI advancements and the digitalization era.

2.2 Case Study

This section presents a comprehensive case study that addresses a significant challenge in the offshore renewable energy sector: the need to develop a life-cycle techno-economic model that incorporates holistic O&M considerations while facilitating spatial assessments of LCoE. First, a literature review on life-cycle techno-economic models is provided in Section 2.2.1. Subsequently, the methodology is explained in Section 2.2.2. Finally, the main results and analysis from the proposed solution are detailed in Section 2.2.3.

2.2.1 Before State-of-the-art

Different life-cycle techno-economic models for offshore renewable energy systems have been proposed in the literature. Among these, techno-economic models based on cost parametric equations have been presented (Castro-Santos et al., 2020; Martinez and Iglesias, 2022a,b; Catalano et al., 2024; Xu et al., 2020). These studies estimate CapEx, covering all pre-operational phases of FOW farm projects, including development and consenting, manufacturing, transmission, and installation stages. However, a common limitation of these life-cycle techno-economic studies is the oversimplified consideration of O&M within the techno-economic framework.

For example, turbine availability is not estimated or modeled in these life-cycle techno-economic models (Castro-Santos et al., 2020; Martinez and Iglesias, 2022a,b; Catalano et al., 2024; Xu et al., 2020). Instead of incorporating specific geographical characteristics, such as metocean conditions and distance to ports, these techno-economic models often use a fixed predetermined turbine availability value (Castro-Santos et al., 2020; Martinez and Iglesias, 2022a,b). Turbine availability is inherently variable depending on the farm site, as it is influenced by local environmental conditions, such as sea states and the proximity to maintenance facilities. Additionally, they frequently neglect to account for turbine availability itself (Catalano et al., 2024; Xu et al., 2020). This oversimplification can lead to inaccurate estimations for O&M operations, particularly concerning turbine downtime. Accurately estimating turbine availability is crucial for realistic operational planning and cost assessments, as even small deviations in availability can have significant impacts on energy production and O&M costs.

Furthermore, in the studies (Castro-Santos et al., 2020; Martinez and Iglesias, 2022a,b; Catalano et al., 2024; Xu et al., 2020), OpEx is estimated deterministically based on a given number of maintenance tasks, without accounting for critical attributes such as reliability, maintainability, accessibility, and availability. Including probabilistic models allows for a more comprehensive assessment by incorporating uncertainties and variability in these factors, offering a more accurate representation of operational costs.

In contrast, several research studies focus exclusively on the O&M phase of FOW farms. Among these studies, the articulation of reliability, maintainability, accessibility, and availability, along with their interdependencies with O&M factors, is mostly achieved through Monte Carlo-based O&M models (Li et al., 2022; Teillant et al., 2012; Dinwoodie et al., 2013; Rinaldi et al., 2018). Monte Carlo-based O&M models use repeated random sampling methods to approximate the failure and repair processes of the FOW farm. However, the main disadvantage of Monte Carlo-based O&M models lies in the high computational burden, because numerous time-consuming simulations are required to achieve convergence in the results (Li et al., 2021). For instance, assessing key performance indicators (KPIs) for a single geographical location can require at least two days of computation (Rinaldi et al., 2018). While these Monte Carlo-based O&M models provide holistic O&M considerations, they lack the computational efficiency needed for broad geospatial assessments. These computations often rely on increasing the number of computational units and employing parallel computing to enhance efficiency. However, this approach imposes high demands on hardware and still fails to deliver sufficient computational efficiency to support large-scale techno-economic analysis and sensitivity analysis. As a result, a novel techno-economic model is required that comprehensively considers O&M factors while facilitating broad spatial assessments of LCoE.

Techno-economic models, both those based on Monte Carlo simulations and those using simpler parametric equations, ensure grounded forecasts by incorporating accurate input data, such as failure rates, weather conditions, and costs, derived from historical records or expert assessments, and by performing verification benchmarking. However, acquiring high-quality input data remains a significant challenge due to its sensitive industrial or commercial nature, underscoring the importance of accurately defining input data from reliable sources, as detailed in Section 2.2.2.

2.2.2 Methodology

The novel methodology for spatially estimating the LCoE focuses specifically on FOW technologies while being adaptable for wave and tidal energy technologies with appropriate modifications. It involves a techno-economic model that includes

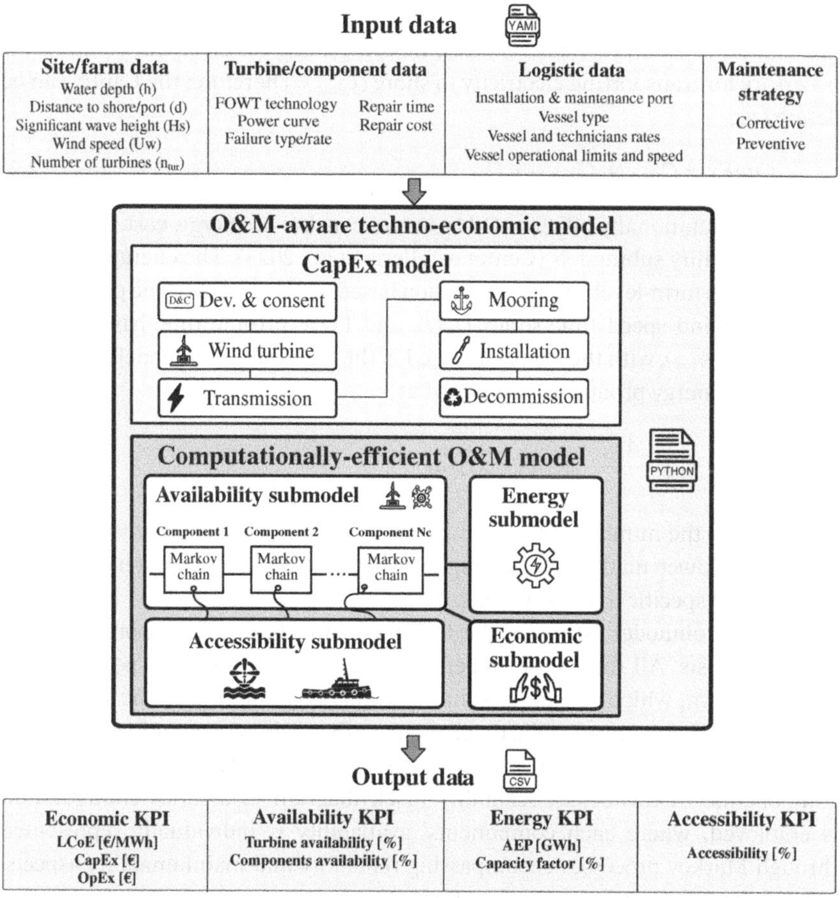

Figure 2.1 The flowchart of the techno-economic model including input and output data.

a CapEx model and a computationally efficient O&M model, which is originally presented in the paper (Centeno-Telleria et al., 2023) and illustrated in Figure 2.1. In this regard, the LCoE is defined as (Centeno-Telleria et al. 2024a),

$$\text{LCoE} = \frac{\sum_{i=1}^{T} (\text{CapEx}+\text{OpEx}_i) \cdot (1+r)^{-i}}{\sum_{i=1}^{T} \text{AEP}_i \cdot (1+r)^{-i}}, \tag{2.1}$$

where AEP_i is the annual energy production, T the operational lifetime of the farm, and r is the discount rate.

Specifically, the CapEx model is a cost model that accounts for the expenses incurred before the operation of an offshore wind farm. These expenses cover development and consent services ($C_{D\&C}$), the manufacturing (C_{manuf}) and the installation of turbines, substructures, moorings, and all electrical components necessary for transporting electricity to shore (C_{inst}). Therefore, the CapEx can be defined as,

$$CapEx = C_{D\&C} + C_{manuf} + C_{inst}. \tag{2.2}$$

The computationally efficient O&M model consists of energy, cost, availability and accessibility submodels (Centeno-Telleria et al., 2023). The energy submodel estimates the farm-level energy production based on the wind turbine power curve ($P(U_w(t))$, wind speed time-series $U_w(t)$, and FOW turbine time-based average availability (A_{tur}), with the latter estimated in the availability submodel. Therefore, the annual energy production is defined as,

$$\text{AEP} = n_{\text{tur}} \cdot \frac{A_{\text{tur}}}{T} \cdot \int_0^T P(U_w(t))\, dt \tag{2.3}$$

where n_{tur} is the number of turbines in the offshore wind farm. For wave energy systems, a power matrix can be employed, while tidal energy systems utilize a power curve specific to tidal systems.

The cost submodel estimates the OpEx as a function of transport, material, and labor costs. All these costs depend on the maintenance tasks performed in the FOW farm, which is estimated in the availability submodel. The availability submodel is the core probabilistic element of the techno-economic model, as it characterizes the failure and repair processes of the FOW turbine and its key components. In this sense, reliability block diagram in a series configuration is employed, where each components' availability is individually represented through Markov processes encompassing reliability and maintainability aspects. A Markov process $X(t)$ is defined by Ross (2014):

- A finite or infinite discrete state space ($\mathcal{X}$), which represents all possible states the process can attain.
- The sojourn time or the time spent in a particular state $i \in \mathcal{X}$ before transitioning to another state follows an exponential distribution.
- A transition probability p_{ij} expressing the probability that, upon leaving state i, $X(t)$ will enter state j, with $\sum_{i,j \in \mathcal{X}} p_{ij} = 1$.
- A transition rate from state i to state j defined as, $\lambda_{ij} = p_{ij} \cdot \lambda_i$.

Finally, the accessibility submodel estimates the waiting time the maintenance crew must wait in port until metocean conditions allow for the maintenance task to be performed as a function of weather windows. In this sense, weather windows are estimated as a function of the vessels' operational limits, the required length of weather windows and metocean conditions as illustrated in Figure 2.2.

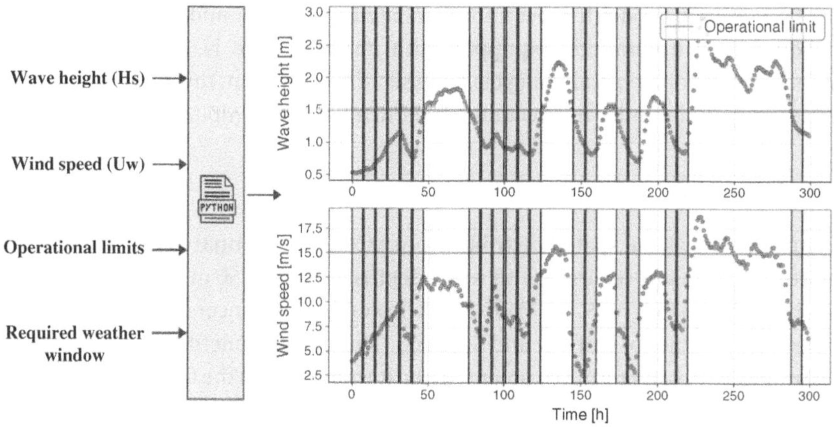

Figure 2.2 Weather windows estimation in the accessibility submodel as a function of metocean conditions, vessels' operational limits and the required weather window.

2.2.2.1 Input Data and Experimental Details

Accurate input data are critical for estimating key metrics in techno-economic models of offshore renewable energy systems. The following input data are utilized in this study:

- Water depth for CapEx modeling is defined from ETOPO Global Relief Model of the NOAA database at one arc-minute resolution provided by Amante and Eakins (2009).
- Significant wave height and wind speed is defined from the ERA5 dataset, which is a comprehensive global climate reanalysis provided by the European Centre for Medium-Range Weather Forecasts (ECMWF) (Hersbach et al., 2020). The data are acquired using the minimum time and spatial resolution available in ERA5, which includes hourly measurements from year 2000 to 2019 and a grid resolution of 0.25° in both longitude and latitude.
- Wind turbine power curve is sourced from the DTU 10-MW Reference Wind Turbine (Bak et al., 2013).
- Data on failure rates, repair times, vessels, and associated costs are defined from multiple sources, including (Centeno-Telleria et al., 2024b; Carroll et al., 2016; IEA Wind TCP Task 49, 2024). These values are critical for accurately modeling the OpEx, as well as turbine availability and, consequently, the overall energy generation.
- The locations of relevant ports are sourced from the World Port Index (National Geospatial-Intelligence Agency, 2023). These data are necessary for determining the logistical costs and transportation times involved in the installation and O&M of offshore wind farms.

The proposed model has been implemented in Python, and simulations were conducted on a computer equipped with an Intel Core i7-1165G7 processor, Windows 10. For a typical life-cycle analysis of a FOW farm, the model completes the evaluation in under five seconds, demonstrating its computational efficiency.

2.2.3 Results

Figure 2.3 shows the spatial distribution of LCoE for potential FOW farms in the North Sea, focusing on areas with water depths exceeding 50 m. The LCoE ranges from approximately €55–75/MWh, with lower values concentrated particularly near the coastlines of Norway and Scotland. These areas benefit from proximity to shore and high wind resource potential, the former reduces the OpEx and the latter increases the AEP. In contrast, higher LCoE values are observed further offshore, where increased distances from shore and more challenging metocean conditions contribute to higher LCoE.

Violin plots are invaluable for visualizing both the central tendency and variability of key metrics. In this context, Figure 2.4 depicts violin plots based on the metrics estimated for each FOW farm site in the North Sea. Specifically, the figure illustrates the density distributions of LCoE, CapEx, OpEx, and capacity factor values.

Figure 2.4 shows the mean LCoE across the North Sea is €64.26/MWh, with CapEx contributing 77.9%, OpEx contributing 22.1%, and a capacity factor being 56.6%. This highlights CapEx as the dominant driver of LCoE. However, the significant OpEx contribution, at over 20%, underscores the importance of reducing

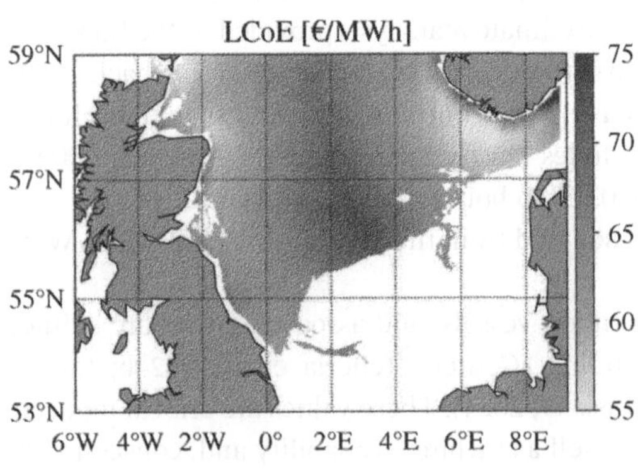

Figure 2.3 The LCoE of potential FOW farms is presented, with locations defined according to water depth greater than 50 m. The points with a more yellow hue represent the areas where offshore wind farm deployment is most cost-effective. (Please refer to the online version for the color image of this figure.)

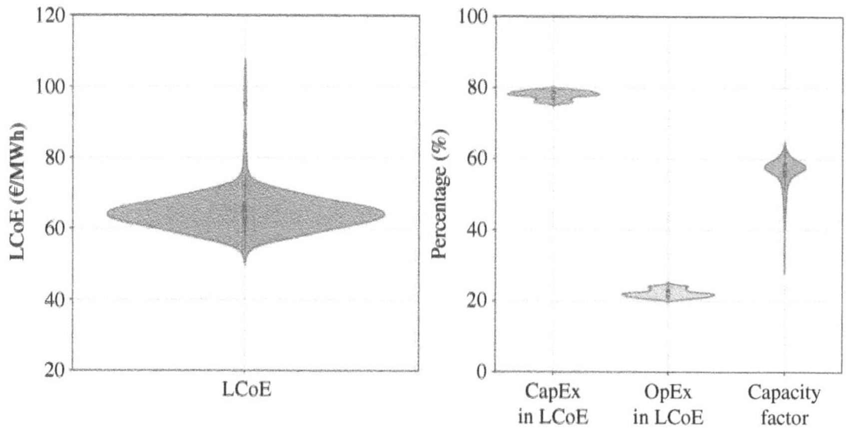

Figure 2.4 Violin plots showing the distribution of *LCoE* (€/MWh) values across the sites evaluated in the North Sea, along with the relative contributions of *CapEx*, *OpEx*, and capacity factor, illustrating how each factor influences the overall cost structure of FOW farms.

Table 2.1 Average values of key metrics for FOW farms evaluated in the North Sea.

Metric	Average value
LCoE (€/MWh)	64.26
CapEx (M€/MW)	3.52
Annual OpEx (k€/MW/year)	55.16
Turbine availability (%)	98.64
Capacity factor (%)	56.62

operational expenditures to achieve further LCoE reductions. In this context, efficient O&M strategies play a pivotal role in minimizing failure rates, turbine downtime, and ultimately lowering OpEx. A key driver for OpEx is the replacement of heavy components, such as blades, gearboxes, and generators, which can contribute to over 75% of the overall OpEx. To summarize, the average values for the main KPIs are summarized in Table 2.1.

2.3 Discussion

The success of the solution proposed in this study can be attributed to several key factors. First, cross-disciplinary collaboration plays a crucial role. The integration of advanced O&M models into techno-economic analysis requires not only knowledge in economics but also expertise in specialized fields such as reliability analysis, advanced simulation and modeling. This collaboration facilitates the transfer of cutting-edge knowledge across disciplines. Moreover, the implementation of this research in the renewable energy industry necessitates close collaboration between academia, the wind industry, and the public sector. By pooling resources, expertise, and real-world data from multiple stakeholders, the collaboration accelerated the technology readiness level (TRL) and its deployment potential.

Without techno-economic models that holistically account for O&M factors, wind farms may have been located based primarily on favorable wind speeds. However, such approaches often overlook critical O&M considerations, where, in addition to wind resources, maintainability is equally crucial. Failing to address these factors could impede the acceleration of the TRL.

This multi-faceted approach not only enhances the accuracy and applicability of the techno-economic framework but also ensures that the solution is grounded in practical realities, fostering its broader adoption. It is important to note that the accuracy and reliability of techno-economic analysis results fundamentally depend on the quality and precision of the input data. Reliable publicly available data can serve as a solid foundation for basic modeling. However, to consider more practical factors with relevant data, which is often accessible only to specific stakeholders, additional shared public datasets or confidential data provided by collaborators are essential. Such data can significantly enhance the credibility and realism of the analysis results, providing a stronger foundation for decision-making and strategic planning.

In addition, it can be anticipated that AI techniques will become the key for both CapEx and OpEx reduction in the future. Throughout different stages of the life cycle of FOW farms, such as design, manufacturing, installation, O&M, and decommissioning, the use of AI has the potential to reduce costs and increase efficiency. For instance, as highlighted in the paper (Badihi et al., 2022; Maldonado-Correa et al., 2020), the AI modeling is a key to realizing lifetime prediction for critical components, e.g. the gearbox, aiming to predict failures in advance and avoid costly replacement. De Anda et al. (2023) suggest that AI techniques can help find the best design solutions for wind turbine structural systems, minimizing mass while maximizing reliability and wind potential. Li et al. (2023, 2024b) quantify the cost reductions in O&M driven by AI-enhanced Prognostics and Health Management (PHM) technologies. Therefore, AI can be employed to address the cost drivers identified through techno-economic models.

A further application of AI involves enhancing the life-cycle techno-economic modeling process itself. By utilizing machine learning algorithms and data analytics, AI can optimize model parameters and identify patterns in historical data, refining the assumptions made during the modeling phase. This approach ultimately improves the overall accuracy of techno-economic models. Among the parameters and assumptions considered, refining the definitions of O&M factors—such as failure rates, repair times, and associated costs—through the application of AI is a key area of focus, due to the inherent uncertainty associated with these factors. Furthermore, advanced maintenance strategies informed by AI algorithms can be integrated into techno-economic models. Existing techno-economic analyses typically rely on conventional corrective or time-based maintenance strategies as inputs. In the era of Industry 4.0–5.0, smart O&M management enabled by cutting-edge technologies such as condition monitoring, predictive analytics, and digital twins holds significant promise for applications in renewable energy systems.

2.4 Conclusion and Future Work

Developing a reliable techno-economic model is crucial for advancing the commercialization of FOW, wave, and tidal energy technologies. This chapter presents a techno-economic model that comprehensively integrates key O&M factors while ensuring computational efficiency for broad spatial LCoE assessments. A case study focused on FOW in the North Sea has been conducted. The North Sea's high wind resource, established infrastructure, and favorable seabed conditions have positioned it as a global leader in offshore wind energy. These characteristics can vary across regions, underscoring the need for site-specific data.

The findings of this study demonstrate the significant wind resource potential of the North Sea, with an average capacity factor of 56.62%. Furthermore, the results, consistent with earlier studies, show that CapEx has the greatest influence on the LCoE of potential FOW farms in the North Sea. However, the OpEx contribution, which exceeds 20%, highlights a significant opportunity for cost reduction and increased competitiveness. In this context, new insights into the distribution of OpEx across the life cycle of FOW farms highlight the emerging role of major component replacements as a critical area for cost reduction. AI can play a vital role in reducing the frequency of these replacements by optimizing O&M strategies, improving predictive maintenance, and enhancing operational decision-making.

Furthermore, the results highlight the computational efficiency of the proposed techno-economic model. For a typical life-cycle analysis of a floating offshore wind farm, the model completes the evaluation in under five seconds, in contrast to the hours required for simulations reported in comparable Monte Carlo-based frameworks (Centeno-Telleria et al., 2023; Rinaldi et al., 2018; Dinwoodie et al., 2013).

The scope of this research focuses on traditional O&M strategies, providing a foundation for future advancements. A key pathway to enhancing cost-effectiveness in FOW farms lies in integrating AI techniques for predictive maintenance. These advanced systems have the potential to transform maintenance planning by reducing OpEx through improved condition monitoring. Nevertheless, the incorporation of AI-driven technologies also increases CapEx. Evaluating trade-offs supports informed decision-making and long-term economic sustainability in offshore renewable energy, where techno-economic models with holistic O&M considerations are essential for effectively balancing costs.

Bibliography

Christopher Amante and Barry W Eakins. Etopo1 arc-minute global relief model: Procedures, data sources and analysis, 2009.

Hamed Badihi, Youmin Zhang, Bin Jiang, Pragasen Pillay, and Subhash Rakheja. A comprehensive review on signal-based and model-based condition monitoring of wind turbines: Fault diagnosis and lifetime prognosis. *Proceedings of the IEEE*, 110(6):754–806, 2022.

Christian Bak, Frederik Zahle, Robert Bitsche, Taeseong Kim, Anders Yde, Lars Christian Henriksen, Morten Hartvig Hansen, Jose Pedro Albergaria Amaral Blasques, Mac Gaunaa, and Anand Natarajan. The DTU 10-MW reference wind turbine. In *Danish Wind Power Research 2013*, 2013.

James Carroll, Alasdair McDonald, and David McMillan. Failure rate, repair time and unscheduled O&M cost analysis of offshore wind turbines. *Wind Energy*, 19(6):1107–1119, 2016.

Laura Castro-Santos, Dina Silva, A Rute Bento, N Salvação, and C Guedes Soares. Economic feasibility of floating offshore wind farms in Portugal. *Ocean Engineering*, 207:107393, 2020.

Matteo Catalano, Idiano D'Adamo, Massimo Gastaldi, and Marzena Smol. An economic analysis of tidal energy to support sustainable development. *World Development Sustainability*, 5:100184, 2024. https://doi.org/10.1016/j.wds.2024.100184.

Manu Centeno-Telleria, Jose Ignacio Aizpurua, and Markel Penalba. Impact of accessibility on O&M of floating offshore wind turbines: Sensitivity of the deployment site. In *Trends in Renewable Energies Offshore*, pp. 847–855. CRC Press, 2022.

Manu Centeno-Telleria, Jose Ignacio Aizpurua, and Markel Penalba. Computationally efficient analytical O&M model for strategic decision-making in offshore renewable energy systems. *Energy*, 285:129374, 2023.

Manu Centeno-Telleria, Hong Yue, James Carrol, Jose I Aizpurua, and Markel Penalba. O&M-aware techno-economic assessment for floating offshore wind farms: A geospatial evaluation off the North Sea and the Iberian Peninsula. *Applied Energy*, 371:123684, 2024a.

Manu Centeno-Telleria, Hong Yue, James Carrol, Markel Penalba, and Jose I Aizpurua. Impact of operations and maintenance on the energy production of floating offshore wind farms across the North Sea and the Iberian Peninsula. *Renewable Energy*, 224:120217, 2024b.

Jonathan De Anda, Sonia E Ruiz, Edén Bojórquez, and Indira Inzunza-Aragon. Towards optimal reliability-based design of wind turbines towers using artificial intelligence. *Engineering Structures*, 294:116778, 2023.

Iain Dinwoodie, David McMillan, Matthew Revie, Iraklis Lazakis, and Yalcin Dalgic. Development of a combined operational and strategic decision support model for offshore wind. *Energy Procedia*, 35:157–166, 2013.

Hans Hersbach, Bill Bell, Paul Berrisford, Shoji Hirahara, András Horányi, Joaquín Muñoz-Sabater, Julien Nicolas, Carole Peubey, Raluca Radu, Dinand Schepers, et al. The ERA5 global reanalysis. *Quarterly Journal of the Royal Meteorological Society*, 146(730):1999–2049, 2020.

IEA Wind TCP Task 49. The IEA Wind Task 49 reference floating wind array design basis. *Technical Report, International Energy Agency (IEA) Wind Technology Collaboration Programme (TCP),* 2024. https://www.ieawind.org/.

Honglin Li, Mingxin Li, James Carroll, and Jie Zhang. Techno-economic analysis incorporating intelligent operation and maintenance management: A case study of an integrated offshore wind and hydrogen energy system. In *Journal of Physics: Conference Series*, pp. 062016, 2024a.

Mingxin Li, Xiaoli Jiang, and Rudy R Negenborn. Opportunistic maintenance for offshore wind farms with multiple-component age-based preventive dispatch. *Ocean Engineering*, 231:109062, 2021.

Mingxin Li, Xiaoli Jiang, James Carroll, and Rudy R Negenborn. A multi-objective maintenance strategy optimization framework for offshore wind farms considering uncertainty. *Applied Energy*, 321:119284, 2022.

Mingxin Li, Xiaoli Jiang, James Carroll, and Rudy R Negenborn. A closed-loop maintenance strategy for offshore wind farms: Incorporating dynamic wind farm states and uncertainty-awareness in decision-making. *Renewable and Sustainable Energy Reviews*, 184:113535, 2023.

Mingxin Li, Xiaoli Jiang, James Carroll, and Rudy R Negenborn. Operation and maintenance management for offshore wind farms integrating inventory control and health information. *Renewable Energy*, 231:120970, 2024b.

Jorge Maldonado-Correa, Sergio Martín-Martínez, Estefanía Artigao, and Emilio Gómez-Lázaro. Using SCADA data for wind turbine condition monitoring: A systematic literature review. *Energies*, 13(12):3132, 2020.

A Martinez and G Iglesias. Site selection of floating offshore wind through the levelised cost of energy: A case study in Ireland. *Energy Conversion and Management*, 266:115802, 2022a.

Abel Martinez and Gregorio Iglesias. Mapping of the levelised cost of energy for floating offshore wind in the European Atlantic. *Renewable and Sustainable Energy Reviews*, 154:111889, 2022b.

National Geospatial-Intelligence Agency. *World Port Index*, 2023. https://msi.nga.mil/Publications/WPI.

G Rinaldi, JCC Portillo, F Khalid, JCC Henriques, PR Thies, LMC Gato, and L Johanning. Multivariate analysis of the reliability, availability, and maintainability characterizations of a spar–buoy wave energy converter farm. *Journal of Ocean Engineering and Marine Energy*, 4:199–215, 2018.

Sheldon M Ross. *Introduction to Probability Models*. Academic Press, 2014.

Boris Teillant, Ronan Costello, Jochem Weber, and John Ringwood. Productivity and economic assessment of wave energy projects through operational simulations. *Renewable Energy*, 48:220–230, 2012.

Xinxin Xu, Bryson Robertson, and Bradley Buckham. A techno-economic approach to wave energy resource assessment and development site identification. *Applied Energy*, 260:114317, 2020. https://doi.org/10.1016/j.apenergy.2019.114317.

3

Making the Most of Data in Offshore Wind Energy: From Population to Physics-informed Modeling

Matthew R. Jones, Daniel J. Pitchforth, Weijiang Lin, Keith Worden, Robin S. Mills, and Elizabeth J. Cross*

Dynamics Research Group, School of Mechanical, Aerospace and Civil Engineering, University of Sheffield, Sheffield, UK

*Corresponding author: Matthew Jones; matthew.r.jones@sheffield.ac.uk

3.1 Introduction

Amid the escalating impacts of climate change, there is an urgent need to accelerate the transition from fossil fuels to renewable energy. Wind is a key renewable energy source, and although the offshore wind sector has seen rapid growth, it is required that the installation rate of offshore wind turbines increases by a factor of four to meet net zero objectives by 2050 (Lee and Zhao, 2022). It is therefore imperative that the cost-effectiveness of the technology continues to improve. Given that the cost of operations and maintenance in offshore wind is notably larger than that of onshore, the implementation of an effective structural health monitoring (SHM) strategy that can inform predictive-based maintenance is crucial for reducing operations and maintenance costs and improving safety.

Over the past decade, the growing development and use of machine learning has led many health monitoring practitioners toward data-driven approaches as potential solutions for models used within predictive maintenance strategies (Avendano-Valencia et al., 2020; Farrar and Worden, 2013; McLean et al., 2023; Rogers et al., 2019). Whilst the flexibility of data-driven learners enables complex, nonlinear relationships to be captured, a widespread transition to the deployment of fully predictive maintenance strategies demands that more progress is made.

Renewable Energy Transition with Artificial Intelligence: Challenge-driven Solutions, First Edition. Nina Dethlefs and Joyjit Chatterjee. Published 2026 by John Wiley & Sons, Inc.

When reliant on machine learners, there is a requirement for training data that sufficiently characterizes the response of the structure across its full operational and environmental envelope, as well as how the structure responds in the presence of damage scenarios of interest. However, it is easy to see how this acts as a significant barrier to the practical rollout of data-driven maintenance strategies. For instance, obtaining datasets representative of the full range of seasonal and operational variation can be challenging, particularly for new assets or where asset owners want to see immediate benefits from a maintenance strategy. Making matters more challenging, sensing data obtained in-situ are often corrupt, sparse, or simply missing, introducing further challenges that standard machine learners are not equipped to handle.

In this chapter, we consider two different paradigms that seek to address limitations with standard data-driven models within the context of structural health monitoring for offshore wind. The first of these is *population-based modeling* (Bull et al., 2021; Gardner et al., 2021; Gosliga et al., 2021; Worden et al., 2015), where rather than considering a single structure in isolation, information and data are shared across multiple structures (a population). The benefit here is twofold; by sharing or pooling data (Bull et al., 2023; Gelman et al., 2013), interdependent relationships between population members can be accounted for and the availability of data can be maximized, with the ultimate aim to improve predictive performance. We apply a population-based approach for modeling the wake effects across a wind farm – an important task for wind turbine health monitoring.

The second paradigm we explore is *physics-informed machine learning* (PIML), which combines physical knowledge with data-based methods to exploit the advantages of both physics-based and data-based modeling philosophies. The former of these looks to mathematically represent a system or process based upon a user's understanding of underlying phenomena, whilst the latter looks to capture relationships directly from measured data. The aim of PIML is to create a flexible model that can provide a degree of insight into its operation and extrapolate effectively within the limits of known physics. One of the major motivating cases for PIML is improving model capabilities when data is scarce, ensuring that the maximum possible performance is achieved from the data available to a user. Here, we explore data scarcity in the context of wave loading predictions, which is fundamental when estimating fatigue accrual and remaining useful life of offshore structures.

The chapter proceeds as follows: Section 3.2 introduces the machine learning theory used in both case studies, which is a form of autoregressive Gaussian process. Section 3.3 discusses the application of population-level modeling for wake modeling, which is applied to the first case study of a simulated wind farm. Section 3.4 introduces physics-informed modeling for wave loading prediction, with the methodology investigated on a second case study consisting of a monopile structure submerged in a wave tank, with a focus on the scenario where there is an absence of suitable data. Finally, Section 3.5 offers some concluding remarks.

3.2 Autoregressive Gaussian Processes

For both of the examples considered in this chapter, autoregressive Gaussian process regression models will be used. Specifically, a Nonlinear AutoRegressive model with eXogenous inputs Gaussian process (GP-NARX). Models with such an architecture are useful when representing variables that are in some way dependent on the past behavior of itself or other variables, which is highly relevant for both capturing the spatial behavior of wind farm wake patterns and temporal correlation in wave loading modeling.

The structure of the autoregressive model enables one to consider additional lagged inputs u_{j-k}, as well as the lagged target y_{j-k}, up to total lags L_u and L_y, respectively. These are then passed through a nonlinear function $f(\mathbf{x})$,

$$y_j = f([u_j, u_{j-1}, ..., u_{j-L_u}, y_{j-1}, y_{j-2}, ..., y_{j-L_y}]) + \varepsilon \tag{3.1}$$

to return a prediction at the current instance y_j. Within the model structure, the nonlinear function here, $f(\mathbf{x})$, is a Gaussian process, with ε a noise term. Using a Gaussian process to model the transition function presents several advantages over NARX models with a fixed functional form, offering flexibility across a wide range of SHM tasks (Gibson et al., 2023; Jones et al., 2023; Lindley et al., 2022) as a result of not being restricted to a fixed basis set. The initial theory provided here is deemed sufficient for the chapter to be self-containing; however, should the reader be interested, further details on GPs are provided in Rasmussen and Williams (2005).

3.3 Population Modeling of Wind Farm Wake Effects

To monitor wind farm performance, the first step is to define a model that is able to characterize the behavior of all turbines within the farm when operating as expected under normal conditions. Such a task exhibits a number of interesting characteristics; firstly, the turbines in a wind farm are of the same model, meaning that they are expected to behave similarly under the same environmental and operational conditions. Secondly, the turbines are positioned within the vicinity of one another, and therefore experience shared environmental conditions. Currently, horizontal-axis wind turbines are the most commonly used, which creates a wake region downstream of each turbine where the wind field is disturbed by the rotating blades. This wake region is generally characterized by decreased mean wind speed and increased turbulence (Archer et al., 2018), and as a consequence, a downwind turbine that is positioned in the wake of an upstream turbine will likely output less power than in free-stream conditions. The reduction in power is often significant, sometimes amounting to a loss of 80% of the free-stream power in compactly spaced wind farms such as Lillgrund (Dahlberg, 2009). Wake effects also influence the mechanical stresses downstream turbines experience, which impact the fatigue loading incurred by a turbine. Understanding wake-induced effects across a wind farm, therefore, is a crucial factor in understanding the expected behavior and the power that a wind turbine will output at a given instance.

The study of wake models can be dated back almost four decades (Lin, 2023), and includes both analytical and numerically-based CFD models. In the case of analytical models, simplified physics laws that govern fluid flow and interactions are used to characterize wake effects, with common examples include the (Jensen, 1983; Frandsen et al., 2006) and Larsen models (Larsen, 1988), all of which are approximate solutions of the Navier–Stokes equations. However, models of this nature are often derived from oversimplifications and idealized conditions, consequently limiting their usefulness. Alternatively, numerical models such as CFD can offer more accurate, higher-fidelity representations of wake behaviors (Janssens and Meyers, 2023; O'Brien et al., 2017). However, they can be very computationally expensive to compute and require validation on real-world data, often rendering their application in industry impractical beyond simple environmental and operational conditions. One potential strategy to alleviate computational bottlenecks is the use of surrogate modeling, where data generated by high-fidelity numerical simulations can be used to learn models that are cheaper and more feasible to run many times (Duthé et al., 2024; Ti et al., 2020). Where measurement data is available from operating turbines, approaches that develop wake models based directly on data obtained from turbines in-situ present a viable solution, with remote sensing technologies such as LiDAR (Renganathan et al., 2022) and SCADA (Nai-Zhi et al., 2022) having been considered. When developing data-driven tools, there is also the opportunity to construct model architectures that reflect the geometry of the wind farm. One example that has become popular when modeling wake effects across wind farms is that of graph neural networks, where the spatial layout of a wind farm is represented as a graph network, which is then fed into a graph neural network (Bentsen et al., (2022; Duthé et al., 2024). Advances here have allowed for improved blade fatigue predictions, as well as the potential for faster prototyping of optimal wind farm layouts.

An alternative framework for deriving a model architecture that reflects the farm geometry that governs the wake behavior is to explicitly consider the wake pattern moving from one turbine to another, which will be the focus of the first case study of this chapter. With the environmental conditions experienced by a given turbine affected by the presence of wakes induced by surrounding turbines, there exists a spatial relationship in the wind conditions throughout the farm that arises from the farm layout. As the incident wind reaches the first-row turbine, the turbine extracts maximal power from the wind and simultaneously produces a downstream wake. For a turbine located in the wake region, output power and its subsequent wake are influenced by the upstream wake conditions, leading to the behavior of turbines along a path of wake progression being spatially autoregressive. This spatial relationship is also impacted by wind direction, resulting in spatial-temporal correlations within the farm.

In this respect, modeling an array of wind turbines can be viewed as a two-level problem: the first addresses the spatial autoregressive wake relationship arising from wake progression; the second is concerned with how the spatial pattern changes with temporally-varying turbine angles. In this work, this task is approached through the use of a Gaussian process-based spatial autoregressive model, which is suitable to capture the nonlinear and spatial autoregressive patterns produced by the wake progression paths.

3.3.1 A Switching GP-SPARX Model

For a set of wake paths, the first level of the model introduced above captures the nonlinear, spatial autoregressive variations using a GP-SPARX model (Lin et al., 2022; Rogers, 2019; Worden et al., 2018),

$$\begin{aligned} u(s,t) &= f\big(u_\infty(t),\ w(i,s)u(i,t)\big), \quad i = 1, \dots, S \\ f(\cdot) &\sim \mathcal{GP}\left(0, k_{SE}(\cdot,\cdot)\right) \end{aligned} \tag{3.2}$$

Here, $u(s, t)$ represents the wind speed at location s and time t. $u_\infty(t)$ is the exogenous input that denotes the free-stream wind speed. $w(i, s)u(i, t)$ is a weighted autoregressive term used to capture the spatial correlation between turbines, where for every turbine i in an array of S turbines, the weight $w(i, s) = 1$ when the output location s is within turbine i's wake, otherwise $w(i, s) = 0$. This definition guarantees that any nonzero term will relate to an upstream turbine that generates wake-shadowing effecting turbine s. Finally, the function $f(\cdot)$ is assumed to be a zero-mean Gaussian process prior with a radial basis function as the kernel.

As the angle of the wind turbine varies, the wake pattern will also change. The proposed solution here is to train a small number of individual GP-SPARX models, where each model relates to a fixed spatial pattern. A higher-level model is then employed to switch between these individual models when applied to test data. In this preliminary study, the selection of patterns used in training and the model switching are performed manually, with the idea that the proposed method can be trialed and examined before developing more complex and automated alternatives.

$$u(s,t) = \begin{cases} f_1\Big(u_\infty(t),\ w(i,s,\phi)u(i,t)\Big) & \underline{\phi}_1 \le \phi < \bar{\phi}_1 \\ \vdots & \\ f_n\Big(u_\infty(t),\ w(i,s,\phi)u(i,t)\Big) & \underline{\phi}_n \le \phi < \bar{\phi}_n \end{cases}, \quad i = 1, \dots, S \tag{3.3}$$

In Eq. (3.3), the output y is acquired by switching between n individual GP-SPARX models $f_1(\cdot), \dots, f_n(\cdot)$, according to the wind direction ϕ. It should be noted that both $f(\cdot)$ and the weight $w(i,s,\phi)$ change with wind direction. $\underline{\phi}$ and $\bar{\phi}$ represent the lower and upper bounds of ϕ, demonstrating that a particular spatial pattern only applies within a given range of wind directions.

3.3.2 A Case Study of a Simulated Wind Farm

To be able to analyze the performance of the proposed model, this study will consider data from a simulated wind farm such that the quality of data can be more readily controlled. Again, future work will consider the challenges that come with the use of operational data.

The simulated wind farm that is used is illustrated in Figure 3.1, where the wake patterns that are associated with the two incident wind directions are highlighted. The dot-dashed lines show a *wake path*, along which the wake propagates from an upstream turbine to a downstream turbine, and is determined by the free-stream wind direction as well as the directions of the local air vortices.

The corresponding mean wind speeds are shown in Figure 3.2. The wind speeds are obtained by using a third-order polynomial NARX model (Lin et al., 2022), providing results that align with commonly known analytical wake models (Archer et al., 2018).

Figure 3.3 illustrates the coverage of the training and testing data with respect to ϕ. In the study presented here, four wake patterns, each associated with a wind direction (indicated by the arrows in Figure 3.3), form the training set used to train four first-level GP models. The testing data covers the full range of wind directions, as shown by the outer circle arrow in Figure 3.3. Predictions on the test set are made using these four first-level models and switching between them at predefined boundaries (dashed lines in Figure 3.3). For simplicity, the boundaries are defined as the midpoints between wind-direction angles used in training.

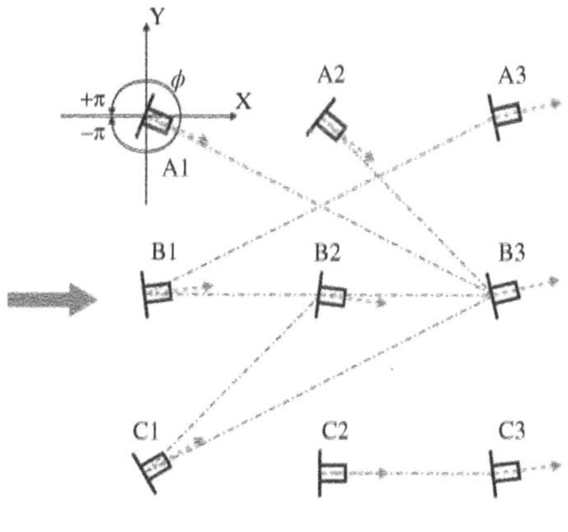

(a) Incoming wind at π rad

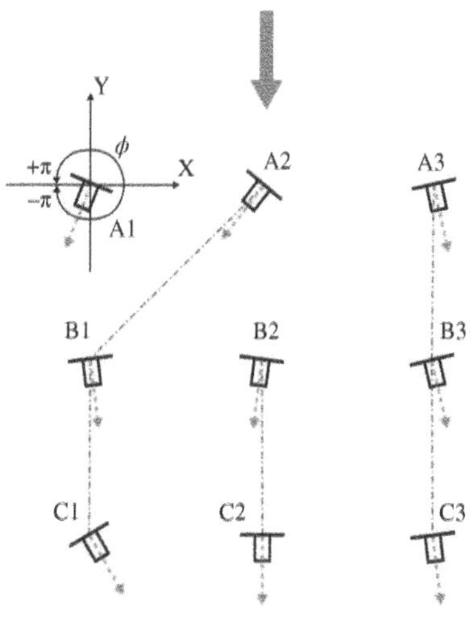

(b) Incoming wind at $\pi/2$ rad

Figure 3.1 Wake patterns in various wind directions in simulated wind farm.

3.3.3 Results

In Lin et al. (2022), it was shown by the authors that a GP-SPARX model can capture nonlinear correlations that exist within a wind farm with reasonable accuracy, offering improved performance compared to NARX and ARX models. To expand on this work, the focus of the case study here is to evaluate the ability of the switching model (the second-level) to predict time-varying spatial patterns across a wind farm.

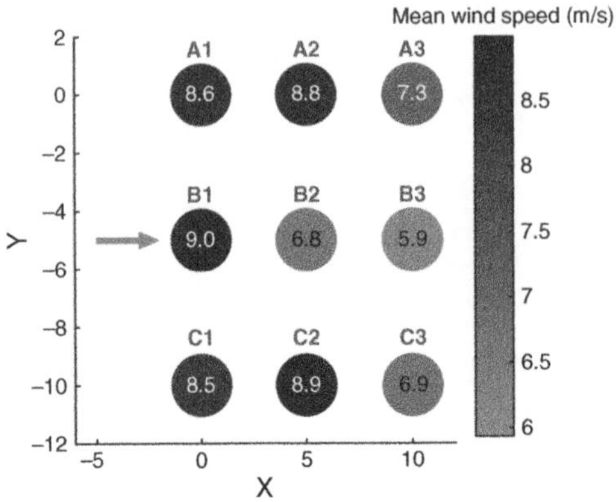

(a) Incoming wind at π rad

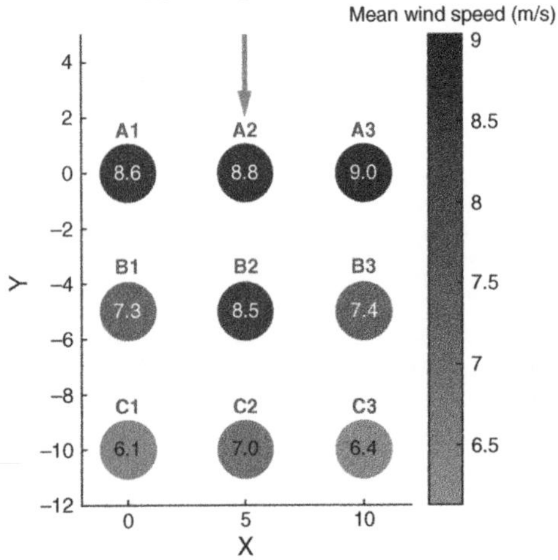

(b) Incoming wind at $\pi/2$ rad

Figure 3.2 Mean wind speed map in various wind directions in simulated wind farm.

Model prediction errors are shown in Figure 3.4, with wind direction represented as the angle, and the distance from the center of the plot showing the squared error at that angle. Two observations can be made: first, that high errors arise in regions that are further from the training data. For predictions at or near training angle (dashes in Figure 3.4), it is seen that the predictions are reasonably accurate, with errors generally reducing the closer that predictions come to the training direction. Away from training angles, the model is forced to extrapolate in terms of wind direction, thus lowering the predictive accuracy. A second observation is that the accuracy of the predictions varies away from the training data, showing spikes of high errors. These spikes indicate that there is potentially a benefit from including wind directions as an input to the model.

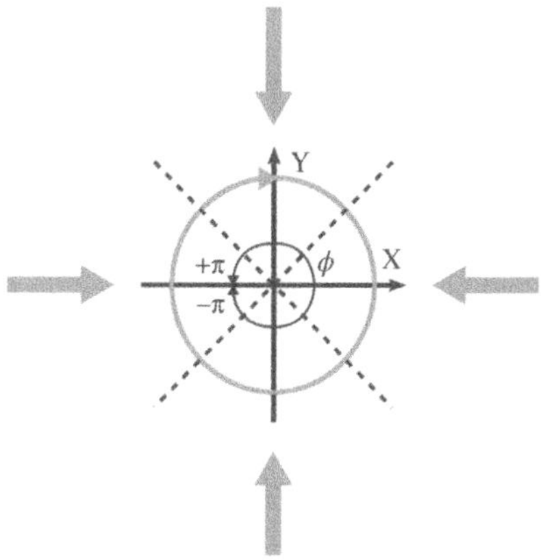

Figure 3.3 Wind directions included in training and test data. Arrows indicate the wind direction used in training data, with the outer circle indicating that all wind directions are used in the testing data. The dotted lines indicate the switching boundaries between models.

3.3.4 Discussion

In this first case study of the chapter, we have presented a population-based method for modeling correlation between individual turbines to capture spatial and temporal variations in wake patterns across a wind farm. The proposed spatial autoregressive model, termed GP-SPARX, demonstrates how one may use the known location of wind turbines to construct an autoregressive structure to model the evolution of wake paths across successive neighboring turbines. Some initial results on a simulated wind farm were presented, demonstrating the potential of the proposed two-level model to be able to predict the spatio-temporal correlations across a wind farm, given only one spatial reference (i.e. weather station measurements), providing a framework for a low-cost indicator of wind farm performance.

3.4 Physics-informed Machine Learning for Wave Loading Prediction

A strength of data-based methods and, by extension, physics-informed machine learners, is the ability to capture relationships directly from data. For scenarios where systems and phenomena are challenging to mathematically describe in complete, the use of measured data can be very useful to aid understanding of the task at hand. The second case study considered in this chapter investigates the combination of measured data and linear wave theory to approximate the wave force acting on a monopile structure.

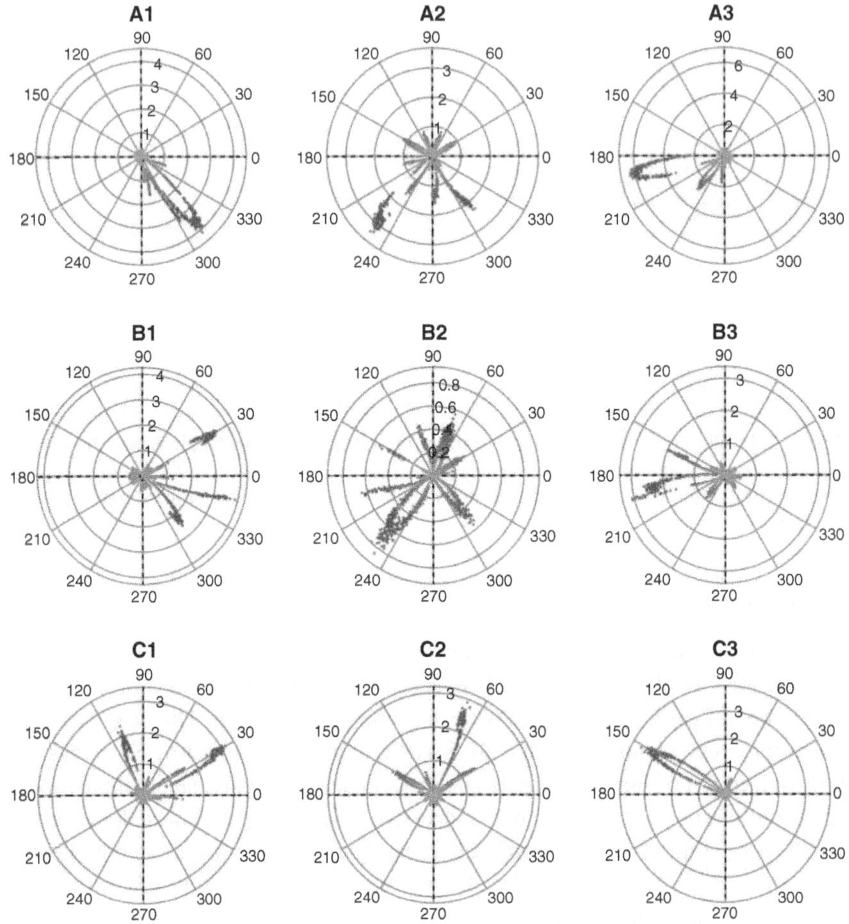

Figure 3.4 Polar map of predictions squared error. Angle represents wind direction and distance from center is squared error of a given prediction. Blue colors indicate lower squared errors, with red representing higher errors. (Please refer to the online version for the color image of this figure.)

The estimation of wave forces acting on structures is a complex task, comprising of many factors: the behavior of the waves themselves, the variability of observed waves, and the interaction between the waves and the structure. Although possible, the measurement of wave loads acting on an offshore structure is both challenging and expensive. During the Christchurch Bay Tower (CBT) (Najafian et al., 2000) project, a bespoke set of force collars was installed on a full-scale structure to measure wave loads in a real sea environment. The force collars were an effective way to capture wave forces, however, by their nature (comprising both moving parts and strain gauges), they were relatively delicate sensors for an offshore environment. A common goal of engineering regression tasks, and the aim of the work outlined below, is to utilize the outputs of more durable, widely available sensors to estimate the outputs of more expensive ones. Here, the incoming wave height, commonly measured by wave buoys and wave radars (Ewans et al., 2014; Tucker, 1991), is used as an input to the models.

Physics-informed machine learning (PIML) combines physical knowledge with data-based methods to benefit from the advantages of either approach used independently. Successful application of PIML can create flexible models that are able extrapolate within the limits of known physics. The means by which combined model structures are created presents an interesting and emerging research topic, overviews of which can be found in Cross et al. (2022) and von Rueden et al. (2021). Within the wave loading community, Pena and Huang (2021) used a CFD model in order to generate training data for a Generative Adversarial Network (GAN). The resulting model achieved comparable performance to the CFD simulation with significant reductions in runtime. Previous work of the authors (Pitchforth et al., 2021) combined Morison's Equation with Gaussian process NARX models to increase predictive performance in cases of limited training data.

The following sections present a case study demonstrating how physical knowledge may be used to assist purely data-based machine learners for wave loading prediction. For context, an overview of the experiment used to generate the data for the study is first shown. Methodology, including important differences in model structure are highlighted, before a discussion of key results.

3.4.1 Monopile Wave Tank Experiment

A series of experiments was performed on a monopile structure within the wave tank at the Laboratory for Verification and Validation (LVV) with the aim of studying wave loading on offshore structures. The controllable, repeatable environment, in which wave conditions could be defined by spectra, was ideal for model development and testing. An ability to mount data logging equipment and hardware alongside the tank, rather than having to attach everything to the structure itself, also allowed for the installation of a dense sensor network. Alongside the wave gauges and force collars required for this work, accelerometers, strain gauges and flowmeters were also installed. The final dataset contained 27 channels, measured at 2048 Hz.

An overview of the experimental setup is shown in the diagram in Figure 3.5, with a photo of the experimental setup shown in Figure 3.6. For additional information on the experiment design, the reader is directed toward (Pitchforth et al., 2023).

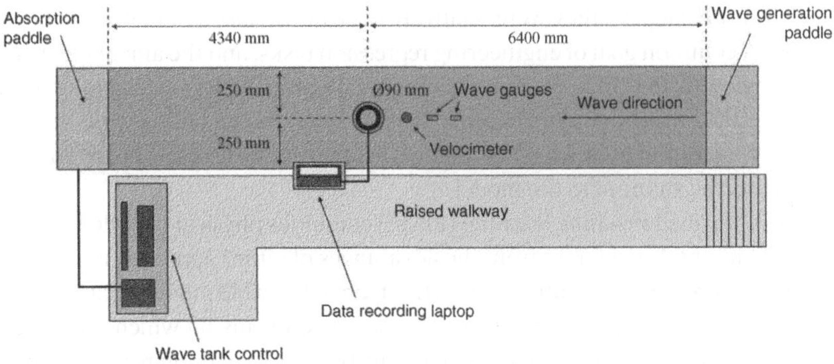

Figure 3.5 Diagram of the experimental setup within the wave tank.

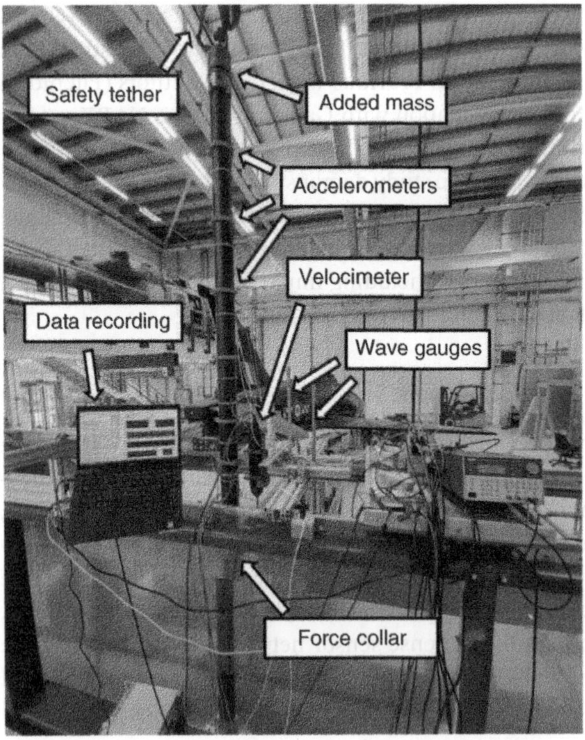

Figure 3.6 Monopile structure and measurement equipment setup within the wave tank.

3.4.2 Model Structure

For any given modeling scenario, the most appropriate choice of physics-informed model structure will vary according to the available physical knowledge and type of data-based learner. Here, the model is required to predict wave load using incoming wave height as an input. As the relationship between these two variables is complex, not only is a flexible model required, but also a sufficient amount of training data to sufficiently capture the complex relationship. To reduce the complexity of the learning task, linear wave theory is called upon to approximate the flow conditions near to the monopile, reducing the complexity of the learning tasks and aiming to increase performance at lower quantities of data. An overview of the model structure is shown Figure 3.7.

As in the previous case study, we consider the use of an autoregressive form of Gaussian process regression, as the data-based component of the model. As the interest is now in predicting wave loading over time, we now switch to a temporal version of GP-NARX, where current time instances (and potentially other

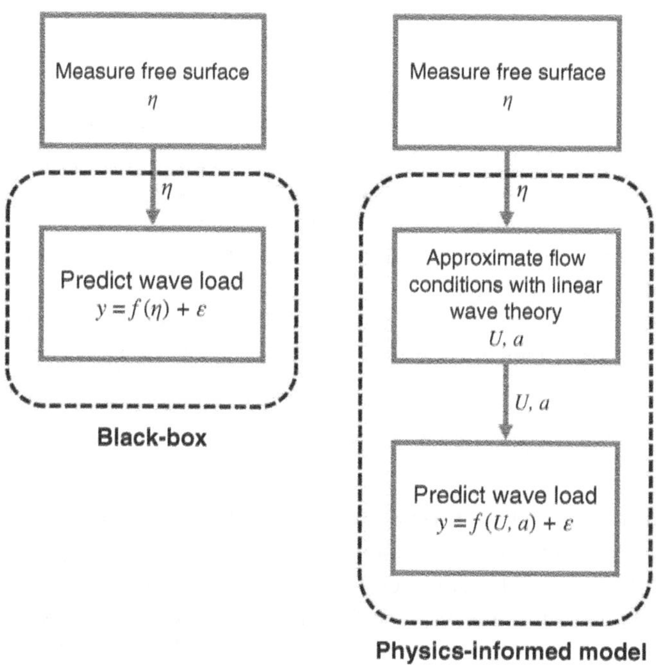

Figure 3.7 Black-box model structure: The measured free surface η is directly used as an input to the GP-NARX. Physics-informed model: Linear wave theory is used, along with η, to approximate flow conditions (water particle velocity U and acceleration a), which are instead used as inputs to the model.

variables) are dependent on past states. A temporal autoregressive architecture is relevant across many wave loading modeling scenarios, including capturing periodic trends, e.g. dominant frequencies and periodicity of waves; capturing a delay, e.g. upstream measurements; and representation on higher order terms, e.g. approximation of phenomena such as vortex shedding (Worden et al., 2017; 2018).

Applying the general form of Eq. (3.1) more specifically to this modeling task gives,

$$y_t = f([u_t, u_{t-1}, ..., u_{t-L_u}, y_{t-1}, y_{t-2}, ..., y_{t-L_y}]) + \varepsilon \tag{3.4}$$

where the previous signal values, y_{t-i} are the wave force and the exogenous inputs u_{t-i} are data from other sensors. Here, for the purely data-based models, u_{t-i} will be measured wave gauge data, whilst for the physics-informed models, u_{t-i} will be the approximated flow conditions. The determination of maximum lags L_u and L_y will affect model structure and performance; they should be selected via an appropriate lag selection process (Hafiz et al., 2018; Paananen et al., 2019; Pitchforth et al., 2021). For this work, optimum lags of $L_u = 4$ and $L_y = 0$ were obtained following Pitchforth et al. (2021).

To obtain an estimation of the wave force from the measured free surface, the methodology is summarized as follows:

1. Retrieve a sum of linear waves for a given wave state via fitting data to free surface measurements from a wave gauge. Check the performance of free surface reconstruction using unseen data from a second wave gauge.
2. Approximate the water particle velocity U and acceleration a close to the monopile, using the expressions derived for the obtained sum of linear waves (Lighthill, 1978; Sarpkaya, 2010).
3. Use the approximated flow conditions as inputs to the GP-NARX to estimate wave force. Test performance using unseen measurements from the force collar.

3.4.3 Results

Models were tested across a matrix of JONSWAP (Hasselmann and Olbers, 1973) waves, defined by a range of peak enhancement factors $\gamma = 1.3 : 1 : 5.3$ and peak frequencies $\omega_p = 0.7 : 0.1 : 1.1$Hz. For each wave state, 300 data points were shown to the model during training, with performance measured on an unseen test set of 1000 points. For this study, the data was downsampled from 2048 to 16 Hz; this allowed a larger time duration of each wave state to be used as training data within the GP-NARX for a given quantity of data. For the frequencies of waves generated in the tank, mostly below 1 Hz, a sample rate of 16 Hz was sufficient.

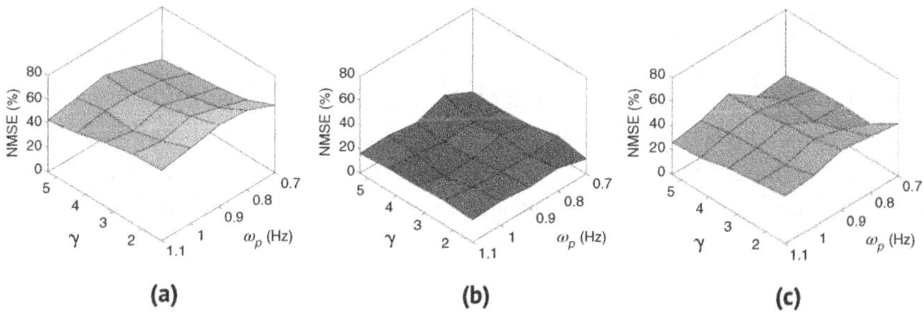

Figure 3.8 NMSE values for a matrix of JONSWAP wave states defined by peak enhancement factor γ and peak frequency ω_p. (a) Black-box: GP-NARX with an SE kernel using wave gauge data as an input. (b) Physics-informed: GP-NARX with an SE kernel using approximated flow conditions as an input. (c) The improvements of the physics-informed model over the black-box model.

Two competing model structures, outlined in Figure 3.7, were compared to assess the benefit of incorporating linear wave theory in the GP-NARX. The performance of the black-box and physics-informed models for a matrix of wave conditions is shown in Figure 3.8.

A clear improvement in performance is observed by the physics-informed model over the purely data-based approach. The use of linear wave theory to approximate flow conditions for use as inputs to the GP-NARX achieved a 15.93% average NMSE across all wave states, compared with 51.37% average NMSE when the measured free surface was passed directly to the GP-NARX. The physics-informed model was also able to outperform the purely data-based approach on all individual wave states, with even the poorest performing wave state (22.34% NMSE for a JONSWAP wave with $\gamma = 5.3$ and $\omega_p = 0.8$Hz) better than the best performing purely data-based model wave state (36.55% NMSE for a JONSWAP wave with $\gamma = 5.3$ and $\omega_p = 0.7$Hz).

An interesting insight drawn from this work, applicable to areas outside wave loading prediction, is that even approximate physical knowledge may assist very flexible machine learning methods. Linear wave theory is known to be an approximation of true underlying fluid behavior (Lighthill, 1978; Sarpkaya, 2010), and it is reliant on a number of strict assumptions. However, despite this, it has been able to offer significant assistance to a GP-NARX, a capable autoregression technique (Worden et al., 2017). For a fixed quantity of available data (300 data points), the GP-NARX was better able to capture the relationship between the wave force and approximated flow conditions than the measured free surface. For many modeling scenarios, extracting as much value as possible from available resources (measured data, existing sensor networks, knowledge of engineers) is a key driver for PIML.

3.5 Conclusions

This chapter has presented two different strategies for maximizing predictive capabilities when developing models for offshore predictive maintenance. First, it was shown that by adopting a population-level perspective of a wind farm, it is possible to form an autoregressive model for capturing the spatial-temporal behavior of the wake effect across a farm. By approaching the problem as a population instead of many independent turbines, the spatial lag structure of the wake path can be identified, as well as an ability to explicitly account for inter-turbine correlations that arise from the layout of the farm. Future work will apply the developed methodology to operational data, where it is likely that a more continuous switch between the different model regimes would be beneficial. One potential solution to be pursued here is the use of a mixture of experts framework (Yuan and Neubauer, 2008), where the overall predictions are a weighted sum of individual GP-SPARX predictions, effectively acting as a gradual switch between different regimes. An additional challenge likely to be faced with operational data is that of extreme and rare operational/environmental conditions. Whilst pooling or sharing data within a population will maximize the conditions observed in the training dataset, to more effectively handle anomalous conditions, robust GP regression methods (Shah et al., 2014) offer a potential strategy to account for outlying behavior. Secondly, physics-informed machine learning was presented as a solution for the estimation of wave loads acting on a monopile structure. Crucially, it is shown that by incorporating linear wave theory into a machine learner, it is possible to retrieve loading predictions from only measurements that would be commonly available on an offshore structure, such as a wave radar, thus significantly increasing the practicality of virtual fatigue estimation. An interesting future avenue will be the integration of higher-fidelity physical models for predicting flow conditions, particularly where the wave fields become more complex and nonlinear, such as when considering fluid-structure interactions.

Acknowledgments

The authors would like to acknowledge the support of the EPSRC, through grant reference numbers EP/R004900/1, EP/S001565/1, and EP/R003645/1, and Innovate UK through the OLLGA project grant 10040817. Finally, MRJ would like to acknowledge the University of Sheffield for support through a Research Excellence Fellowship.

Bibliography

CL Archer, A Vasel-Be-Hagh, C Yan, S Wu, Y Pan, JF Brodie, and AE Maguire. Review and evaluation of wake loss models for wind energy applications. *Applied Energy*, 226:1187–1207, 2018.

S Ashwin Renganathan, Romit Maulik, Stefano Letizia, and Giacomo Valerio Iungo. Data-driven wind turbine wake modeling via probabilistic machine learning. *Neural Computing and Applications*, 34(8):6171–6186, 2022.

Luis David Avendano-Valencia, Eleni N Chatzi, and Dmitri Tcherniak. Gaussian process models for mitigation of operational variability in the structural health monitoring of wind turbines. *Mechanical Systems and Signal Processing*, 142:106686, 2020.

Lars Ødegaard Bentsen, Narada Dilp Warakagoda, Roy Stenbro, and Paal Engelstad. Wind park power prediction: Attention-based graph networks and deep learning to capture wake losses. In *Journal of Physics: Conference Series*, volume 2265, pp. 022035. IOP Publishing, 2022.

Lawrence A Bull, Paul A Gardner, Julian Gosliga, Timothy J Rogers, Nikolaos Dervilis, Elizabeth J Cross, Evangelos Papatheou, AE Maguire, Carles Campos, and Keith Worden. Foundations of population-based SHM, Part I: Homogeneous populations and forms. *Mechanical Systems and Signal Processing*, 148:107141, 2021.

Lawrence A Bull, Matthew R Jones, Elizabeth J Cross, Andrew Duncan, and Mark Girolami. Encoding domain expertise into multilevel models for source location. *arXiv preprint arXiv:2305.08657*, 2023.

EJ Cross, SJ Gibson, MR Jones, DJ Pitchforth, S Zhang, and TJ Rogers. Physics-informed machine learning for structural health monitoring. In *Structural Health Monitoring Based on Data Science Techniques*, pp. 347–367. Springer, 2022.

J-A Dahlberg. Assessment of the lillgrund windfarm: Power performance and wake effects. Technical report, Vattenfall Vindkraft AB, 2009.

Gregory Duthé, Francisco de N Santos, Imad Abdallah, Wout Weijtjens, Christof Devriendt, and Eleni Chatzi. Flexible multi-fidelity framework for load estimation of wind farms through graph neural networks and transfer learning. *Data-Centric Engineering*, 5:e29, 2024.

Kevin Ewans, Graham Feld, and Philip Jonathan. On wave radar measurement. *Ocean Dynamics*, 64:1281–1303, 2014.

CR Farrar and K Worden. *Structural Health Monitoring: A Machine Learning Perspective*. John Wiley and Sons Ltd, 2013.

Sten Frandsen, Rebecca Barthelmie, Sara Pryor, Ole Rathmann, Søren Larsen, Jørgen Højstrup, and Morten Thøgersen. Analytical modelling of wind speed deficit in large offshore wind farms. *Wind Energy: An International Journal for Progress and Applications in Wind Power Conversion Technology*, 9(1-2):39–53, 2006.

P Gardner, LA Bull, J Gosliga, N Dervilis, and K Worden. Foundations of population-based SHM, Part III: Heterogeneous populations–Mapping and transfer. *Mechanical Systems and Signal Processing*, 149:107142, 2021.

Andrew Gelman, John B Carlin, Hal S Stern, David B Dunson, Aki Vehtari, and Donald B Rubin. *Bayesian Data Analysis*. CRC Press, 2013.

Samuel J Gibson, Timothy J Rogers, and Elizabeth J Cross. Distributions of fatigue damage from data-driven strain prediction using Gaussian process regression. *Structural Health Monitoring*, 22(5):3065–3076, 2023.

J Gosliga, PA Gardner, LA Bull, N Dervilis, and K Worden. Foundations of population-based SHM, Part II: Heterogeneous populations–Graphs, networks, and communities. *Mechanical Systems and Signal Processing*, 148:107144, 2021.

Faizal Hafiz, Akshya Swain, Eduardo MAM Mendes, and Nitish Patel. Structure selection of polynomial NARX models using two dimensional (2D) particle swarms. In *2018 IEEE Congress on Evolutionary Computation (CEC)*, pp. 1–8, 2018.

K Hasselmann and Dirk Olbers. Measurements of wind-wave growth and swell decay during the Joint North Sea Wave Project (JONSWAP). *Ergänzung zur Deut. Hydrogr. Zeitschrift, Reihe A (8)*, 12:1–95, 1973.

Nick Janssens and Johan Meyers. Towards real-time optimal control of wind farms using large-Eddy simulations. *Wind Energy Science Discussions*, 2023:1–42, 2023.

NO Jensen. *A Note on Wind Generator Interaction*. Risø National Laboratory. Risø-M No. 2411,1983.

MR Jones, TJ Rogers, and EJ Cross. Constraining Gaussian processes for physics-informed acoustic emission mapping. *Mechanical Systems and Signal Processing*, 188:109984, 2023.

GC Larsen. *A Simple Wake Calculation Procedure*. Risø National Laboratory. Risø-M No. 2760,1988.

J Lee and F Zhao. Global wind report 2022. Technical report, Global Wind Energy Council, 2022.

James Lighthill. *Waves in Fluids*. Cambridge University Press, 1978.

Weijiang Lin. Monitoring a population of structures from a spatiotemporal perspective: An application to offshore wind farms. PhD thesis, University of Sheffield, 2023.

W Lin, K Worden, and EJ Cross. A spatial autoregressive approach for wake field prediction across a wind farm. In *Proceedings of EWSHM 2022 – European Workshop on Structural Health Monitoring*, volume 270, pp. 530–540. Palermo, Italy: Springer Nature, 2022.

CA Lindley, S Beamish, RS Dwyer-Joyce, N Dervilis, and K Worden. A Bayesian approach for shaft centre localisation in journal bearings. *Mechanical Systems and Signal Processing*, 174:109021, 2022.

Jacques H Mclean, Matthew R Jones, Brandon J O'Connell, Eoghan Maguire, and Tim J Rogers. Physically meaningful uncertainty quantification in probabilistic wind turbine power curve models as a damage-sensitive feature. *Structural Health Monitoring*, 22(6):3623–3636, 2023.

Guo Nai-Zhi, Zhang Ming-Ming, and Li Bo. A data-driven analytical model for wind turbine wakes using machine learning method. *Energy Conversion and Management*, 252:115130, 2022.

G Najafian, RG. Tickell, R Burrows, and JR. Bishop. The UK Christchurch Bay Compliant Cylinder Project: Analysis and interpretation of Morison wave force and response data. *Applied Ocean Research*, 22:129–153, 2000.

JM O'Brien, TM Young, DC O'Mahoney, and PC Griffin. Horizontal axis wind turbine research: A review of commercial CFD, FE codes and experimental practices. *Progress in Aerospace Sciences*, 92:1–24, 2017.

Topi Paananen, Juho Piironen, Michael Riis Andersen, and Aki Vehtari. Variable selection for Gaussian processes via sensitivity analysis of the posterior predictive distribution. In *Proceedings of the 22nd International Conference on Artificial Intelligence and Statistics*, pp. 1743–1752. Proceedings of Machine Learning Research, 2019.

Blanca Pena and Luofeng Huang. Wave-GAN: A deep learning approach for the prediction of nonlinear regular wave loads and run-up on a fixed cylinder. *Coastal Engineering*, 167:103902, 2021.

DJ Pitchforth, TJ Rogers, UT Tygesen, and EJ Cross. Grey-box models for wave loading prediction. *Mechanical Systems and Signal Processing*, 159:107741, 2021.

Daniel J Pitchforth, Robin S Mills, Timothy J Rogers, Ulf T Tygesen, and Elizabeth J Cross. Physics-informed Gaussian processes for wave loading prediction. *Structural Health Monitoring 2023*, 2023.

Carl Edward Rasmussen and Christopher KI Williams. *Gaussian Processes for Machine Learning*. The MIT Press, 2005.

TJ Rogers. Towards Bayesian system identification: With application to SHM of offshore structures. PhD thesis, Department of Mechanical Engineering, University of Sheffield, 2019.

TJ Rogers, K Worden, R Fuentes, N Dervilis, UT Tygesen, and EJ Cross. A Bayesian non-parametric clustering approach for semi-supervised structural health monitoring. *Mechanical Systems and Signal Processing*, 119:100–119, 2019.

Turgut 'Sarp' Sarpkaya. *Wave Forces on Offshore Structures*. Cambridge University Press, 2010.

Amar Shah, Andrew Wilson, and Zoubin Ghahramani. Student-t processes as alternatives to Gaussian processes. In *Artificial Intelligence and Statistics*, pp. 877–885. PMLR, 2014.

Zilong Ti, Xiao Wei Deng, and Hongxing Yang. Wake modeling of wind turbines using machine learning. *Applied Energy*, 257:114025, 2020.

MJ Tucker. *Waves in Ocean Engineering: Measurement, Analysis, Interpretation*. Ellis Horwood, 1991.

Laura von Rueden, Sebastian Mayer, Katharina Beckh, Bogdan Georgiev, Sven Giesselbach, Raoul Heese, Birgit Kirsch, Michal Walczak, Julius Pfrommer, Annika Pick, et al. Informed machine learning – a taxonomy and survey of integrating prior knowledge into learning systems. *IEEE Transactions on Knowledge and Data Engineering*, 35(1):614–633, 2021.

K Worden, William Becker, TJ Rogers, and EJ Cross. On the confidence bounds of Gaussian process NARX models and their higher-order frequency response functions. *Mechanical Systems and Signal Processing*, 104:188–223, 2018.

K Worden, EJ Cross, N Dervilis, E Papatheou, and I Antoniadou. Structural health monitoring: From structures to systems-of-systems. *IFAC*, 48:1–17, 2015.

K Worden, TJ Rogers, and EJ Cross. Identification of nonlinear wave forces using Gaussian process NARX models. In *Nonlinear Dynamics*, volume 1, pp. 203–221, 2017.

Chao Yuan and Claus Neubauer. Variational mixture of Gaussian process experts. *Advances in Neural Information Processing Systems*, volume 21, pp. 1897–1904, 2008.

4

Leveraging the Power of Informal Networks in Renewables

Rakesh Maharaj[1] *and James Akrigg*[2]

[1]*ARMSA Academy, Manchester, UK*
[2]*ARMSA Academy/Cloudy Software Limited, Manchester, UK*

4.1 Challenge

As 2050 gets ever closer, the global warming target of 1.5° is at risk. The transition to renewable energy still offers hope in the race to net zero, but only if the sector can overcome barriers and scale up at an unprecedented pace. To do this effectively, efficiently, and safely, renewables businesses need to rethink how they approach organizational agility and build operational resilience.

The background to the sector is well known. As it undergoes unprecedented rapid expansion, it is characterized by a growing skills gap; inexperienced staff working in unfamiliar environments; high staff churn; a focus on traditional approaches to training and development that emphasize technical knowledge, rather than decision-making capabilities; and rapid technological innovation, which presents unknown risks.

These characteristics create a sector where operational demand is always significantly outstripping organizational capacity. And where there is such a gross disproportion between demand and capacity, there will always be under-delivery problems, delays, breakdowns, and safety issues. Organizational capacity in terms of people is clearly not just about increasing headcount; it is the ability for people to collectively approach each situation critically and thoughtfully, so that, at a decentralized level, they can make decisions in an adaptive and resilient way.

Within this context is another layer of complexity – the rapid movement of people and organizations involved in renewable assets. The organization involved in the development phase may not be the organization that is responsible for construction, which may not be the organization responsible for operation and maintenance. The asset is not only changing hands in terms of ownership and oversight but, at each point in those phases, the supply chain is constantly changing.

With large projects transforming over time and involving new stakeholders, the opportunity for knowledge transfer is limited. So, not only is there a challenge with people and skills, but there is also a challenge presented by changing organizations and perspectives. Alongside the people-orientated issue, there is a clear organization-oriented issue. These factors make it difficult to identify potentially problematic projects before the construction phase, often leaving organizations dealing with the legacy of earlier decisions for years to come.

The ramifications of this are driven home in evidence from a 2024 study (ARMSA Academy, 2024) conducted with professionals in renewable energy companies across Europe. The findings show that 75% of respondents reported that unexpected failures and defects were a significant impact of poor early-stage decision-making or access to information. This indicates that strengthening the connection between early-stage development decisions and their downstream impacts is critical.

Within renewables organizations, a key challenge is how to effect improvement and foster growth in a safe and sustainable way, and ensure the right decisions are made at the right time. And, in the complex adaptive organizations that typify the sector, the answer is not to impose singular top-down management systems because such management systems are unlikely to work in fast-paced, dynamic environments. Instead, the focus needs to be on investing in people and skills – building a knowledge economy that helps to address the mismatch between operational demand and organizational capacity, manage the complexities involved, and enable the sector to scale effectively and safely.

In addressing the complex challenges outlined above, and to facilitate a smoother transition and meet global targets, there needs to be a paradigm shift in the sector. Doing more of the same will only worsen issues like construction overruns, equipment failures, and safety incidents.

To close the gap between operational demand and organizational capacity, organizations need to invest in their people and develop cultures that support a learning organization. As characterized by Senge (1990a), learning organizations are places where "people continually expand their capacity to create the results they truly desire, where new and expansive patterns of thinking are nurtured, where collective aspiration is set free, and where people are continually learning to see the whole together."

Senge (1990b) also points out that organizations that will excel are those that "discover how to tap people's commitment and capacity to learn at all levels." Tapping into people's desire to learn, encouraging new thinking and collaboration, and strengthening their ability to make informed decisions are key challenges for all renewables organizations as the sector scales up at an unprecedented rate. And part of the answer is to improve how people learn, collaborate, and are given the skills and confidence to make decisions at every stage of a project lifecycle.

People-centered digital tools, such as the case study tool h'alt®, that support learning and decision-making in the workflow can support skills development and encourage teams to collaborate across functions. By incorporating AI technologies, these tools are further strengthened so they can learn from the questions people ask, analyze the valuable data gathered, and produce organizational insights. This feedback from real-world local data can then be used to identify key leverage points to drive improvement and prevent small problems from becoming bigger problems.

4.2 Case Study

So far, the sector has been slow to recognize the potential of applying AI to support a data-driven, people-centered approach to managing organizational change. Most of the applications of AI in renewables focus on performance optimization of generation plant and associated equipment. And those software applications aimed at enhancing human capital tend to be compliance and process control-based, rather than looking more broadly at enabling people to use agile and critical thinking to drive performance.

Yet, digital decision-making tools, supported and enhanced by AI, can drive an approach to improving performance and resilience that is both people-centric and organization-centric. Organizations are increasingly recognizing that, if they are going to be successful, they can no longer rely on traditional command and control-based management practices.

Bersin (2022, p. 9), for example, talks about how the most effective companies are not hierarchical. Instead, business drivers such as agility, engagement, and empowerment are creating a new organizational model – a network of teams. Bersin (2022, p. 10) explains how these teams "learn and adapt more quickly than individual workers because they challenge each member to develop more creative ideas." And he argues that "as organizations shift their focus to learning, they will increase their reliance on teams to navigate uncertainty and generate growth."

Digital decision-making tools support individual skills development, but they also foster team collaboration and cross-cultural cohesion, giving users confidence in their decision-making and helping to develop high-functioning teams across geographies. And in-workflow learning means people apply what they learn immediately at the point of need to solve problems.

Consuming both tacit and explicit content accelerates the learning process. Tacit content relates to knowledge gained through experience, providing an understanding of "why" (the purpose of their activity) and "what" (the expected outcomes). Explicit content relates to the "how," which is often conveyed through processes. Understanding the "why" and "what" enables the user to question the "how," which is vital if processes are poorly written, generic, or do not suit real-life circumstances. This provides users with the confidence to make decisions about whether to go ahead or not, which is vital in safety-critical activities.

The application of AI increases the ability of decision-making tools to monitor and assess users' patterns of engagement to identify knowledge gaps and other key risks, suggesting content to address these gaps, and producing actionable management insights. The way in which these tools gather data and analyze the organizational intelligence they receive means that necessary improvements can be made earlier in the project lifecycle. These improvements will also be based on what is really happening on the ground and within organizations, rather than management's perception or expectation of what is happening.

Where there are disparate teams operating globally in different contexts and cultures, these tools can provide a cross-portfolio understanding of what the unresolved issues are, so that they can be addressed systemically. Engagement data identifies user needs in context, so local work leaders can spot risks and opportunities, as well as pinpoint high-functioning teams and those needing additional support.

4.2.1 Before SOA: What Was the State-of-the-art/Accepted Solution in the Past?

The renewables sector's current approach to organizational agility and operational resilience remains rooted in custom and practice. There is also a tendency to try to simply transfer processes and practices used in traditional high-hazard industries, such as oil and gas, to offshore wind. But when sectors are at opposite ends of the market maturity cycle – with different skillsets, contracting models, and hazard profiles – you cannot just apply what works in one to the other.

Fast-paced emerging industries cannot afford to fall back on traditional approaches. Where operational demand is outstripping organizational capacity, organizations have to invest in new ways of working and approaching problems if they are to fill that void and rebalance.

Despite this imperative, too many organizations are still being held back by reliance on traditional management systems and reactive workforce planning and skills development. This leaves people on the ground without the requisite knowledge or tools to make agile, informed decisions in uncertain and high-pressure circumstances. The result is asset breakdowns, loss of production, delayed construction, and threats to safety and health. For example, data from G+, the Global Offshore Wind Health and Safety Organisation (2024), which collects health and safety incident data for offshore wind sites across the world, shows that in 2022, there were 225 high-potential incidents, accounting for 26% of all incidents. This fell in 2023 to 183 high-potential incidents, accounting for 11% of the total. There was also a reported fatality in 2023, followed by a further two reported fatalities in 2024.

The International Renewable Energy Agency (IRENA) (2023) identifies "skills and institutional capacity" as a key structural barrier to achieving the COP28 renewable energy targets. And, as a solution, it recommends "awareness- and capacity-building of institutions, communities and individuals to acquire requisite skills and knowledge to drive and sustain the energy transition." IRENA's estimates (2023) suggest that under the 1.5 °C scenario, employment in renewable energy sectors is expected to triple from 2021 levels to around 40 million jobs globally by 2050.

The lack of available skilled staff, compounded by rapidly evolving technologies, means professionals transitioning to renewables are often coming into the industry without the requisite skills, training, and experience. This is particularly true in the new, emerging global markets. At the same time, organizations, teams, and individuals are making decisions at speed and working around differences in national and company cultures.

To tackle the skills gaps in global markets, attract people into the sector, and protect safety and health, organizations must find new ways to upskill people more quickly, effectively, and efficiently. This includes identifying innovative ways to develop and accelerate how people learn the skills they need and supporting them in applying those skills in their job, so they have the confidence and competence to make decisions in uncertain and evolving environments.

If this can be achieved, it should reduce the demands on management to micromanage or answer every question, freeing them up to focus on resolving more complex or challenging issues or queries. This is essential in organizations that are operating with lean teams during the early phases of a project and where they must upscale rapidly to deliver on the project.

One of the key problems in the sector is that investment is overwhelmingly technology-focused, with only limited investment going into developing human capital. And in terms of skills development, instead of innovating, many organizations are continuing to implement conventional forms of learning and personnel development. And, where they are investing, this too often focuses on traditional face-to-face classroom-based training or conventional eLearning packages.

These conventional approaches offer a way of effectively delivering general content in certain contexts. But they tend to be limited by their "push" approach to developing staff. Learners typically receive large volumes of information all in one go and are left to take what they need back to the workplace, relying on their memory to apply it at the point of need.

A push-based approach, however, no longer satisfies the needs of those who design, install, manage, and maintain offshore assets as they work within increasingly challenging, fast-paced, digital environments. In these kinds of environments, which typify the renewables sector, individual, business, and safety performance all rely on informed decision-making to manage risk, find solutions to problems, and identify opportunities.

Safe and effective decision-making relies on accurate, trusted, and detailed information. Professionals dealing with hazardous environments are required to make informed decisions and, to do this, they need access to reliable data. They also need to develop the confidence to make decisions using that data. Whatever work they are involved in – whether contracting a supplier, negotiating prices, hiring staff, or planning work – every decision affects operational control, either by improving or harming performance.

When people are making these decisions, they need content and data that is easily accessible and tells them the potential or likely impacts on the ground, the issues they need to consider, the options available, and the option that best suits the situation. Learning on the job, while doing daily tasks, has long been recognized as a route to professional development. But modern technologies and evolving demands on the workforce have introduced a different dimension, allowing "in-workflow" support to be delivered in innovative ways using people-centered solutions.

The concept of in-workflow support is not new. Bersin (2018) identified "learning in the flow of work" as a new paradigm for training and learning as far back as 2018. What is now changing is the ability to deliver this type of learning using new digital technologies, including AI.

Instead of "push-development," in-workflow digital performance support focuses on "pull-development." And because learners call upon it at the point of need, when they are looking for specific help or have a problem to solve, they find the content they receive very real and relevant. This makes them more likely to pay attention to the information, and to use what they are learning as they make critical decisions on the ground.

Digital decision-making tools, accessed at the point of need within the workflow, both inform renewables professionals' decision-making and help them to appreciate and understand the foreseeable outcomes of their day-to-day decisions. This helps to minimize risk and optimize the performance of both individuals and organizations.

4.2.2 Influencing and Educating Informal Networks

Another key issue in the industry is over-reliance on custom and practice, regulatory approaches, and imposing traditional management systems to effect improvement. These generally result in a top-down, hierarchical approach, based on repeatable steps, and formalized procedures and policies. Like their colleagues in other sectors, senior executives in renewables organizations are adopting these systems as the answer to the challenges they are facing and in the hope that they will improve performance.

Traditional management systems can work well for some organizations, but the renewables sector is not an established, mature sector. The speed with which it is growing and developing, its contracting model and its hazard profile, coupled with constantly evolving technologies, means it exhibits certain characteristics that do not fit a traditional management systems approach. When those further down the management structure try to deliver this approach, they become frustrated that they cannot proceduralize everything because, as soon as they write a procedure, everything changes again. The dynamic nature of the sector, with large projects constantly transforming over time and bringing in new stakeholders, also means success starts to rely upon the degree to which several different management systems (those of the clients and the various supply chains) align.

Related to this is the need for organizations to recognize that the real power of decision-making rarely lies within the formalized organization, but within informal networks, and these do not work according to management system-based corporate procedures or written compliance arrangements. If decision-making relies primarily on informal networks, this means those decisions heavily depend on personnel competencies, capabilities, and experience. Organizations, therefore, have two options. They can either continue to force procedural adherence onto people, knowing the informal network will always exist. Or they can harness the power within those informal networks by influencing and educating the network.

This is where AI-enabled digital decision-making tools can be effective by tapping into and strengthening informal networks. Through embedding knowledge and creating a common set of vocabularies and terms within teams, such tools equip people at all levels, so that collective decisions in the informal network become informed. Developing individual skills is not sufficient – there needs to be collaboration, information transfer between people and teams, and sharing of experiences, including successes and failures.

4.2.3 Methodology

As discussed, digital decision-making tools, supported and enhanced by AI, can drive a people-centric approach to improving organizational performance and resilience, laying the foundation for a more sustainable future.

The team initially developed h'alt® as a decision-making tool for the offshore wind industry, but it is easily adaptable and applicable across the renewables sector. The first step was to undertake extensive market research (ARMSA Academy, 2020), including talking to industry professionals to find out more about the sector's current challenges and aspirations. Participants were asked how they develop decision-making capabilities to optimize performance and minimize risk, and what they looked for in terms of support.

One third of respondents looked for critical thinking and stewardship within their departments. Half the professionals also looked for strong cross-departmental collaboration with a focus on asset reliability and integrity. Safety, procurement, and contracts were identified as areas where cross-collaboration is most critical. All respondents felt that training and development focused on project planning and preparedness can reduce operating costs in the medium term and construction costs in the short term.

The next step was to study key roles within the wind power market to identify individual and collective decisions that affect organizational effectiveness, plant efficiency, and operational safety. Based on the findings, the team's subject matter experts developed h'alt®'s initial 120 video-based "performance sprints" lasting around five to seven minutes, and each linked to the roles and decisions identified. The content, which is biased toward front-end loaded decision-making, is based on real-world investigations, audits, and transformation projects.

Over 18 months, the team conducted pilot testing to ensure the content met user needs. The aim was to devise clear methods of engagement that resonated with users. This meant the user interface had to be familiar and engaging and, from a learning perspective, short-form content is vital. The content also had to be authentic, trusted, and authoritative, delivered by industry experts.

The tool embraces a flexible approach to content development that depends on an organization's needs. Clients can use entirely "sprint" content, choose to develop their own content and connect it to h'alt®, or content developers can build new bespoke content to client specification.

Users "pull" the learning they are looking for by submitting their search queries via a communications platform such as Microsoft Teams. When h'alt®'s machine learning-enabled mentor chatbot receives a query, it retrieves and recommends content, either from the sprints or the organization's own content. It then offers this learning to the user to support their workplace query. If the user searches for content that is not available, the mentor flags the query as unresolved, and h'alt® points the individual directly to their manager if the unresolved request is critical. Non-critical unresolved requests are then addressed by managers during monthly team briefs.

Figure 4.1 shows how users work with h'alt® to support their decision making. Users interact with h'alt® on their Microsoft Teams application in two main ways.

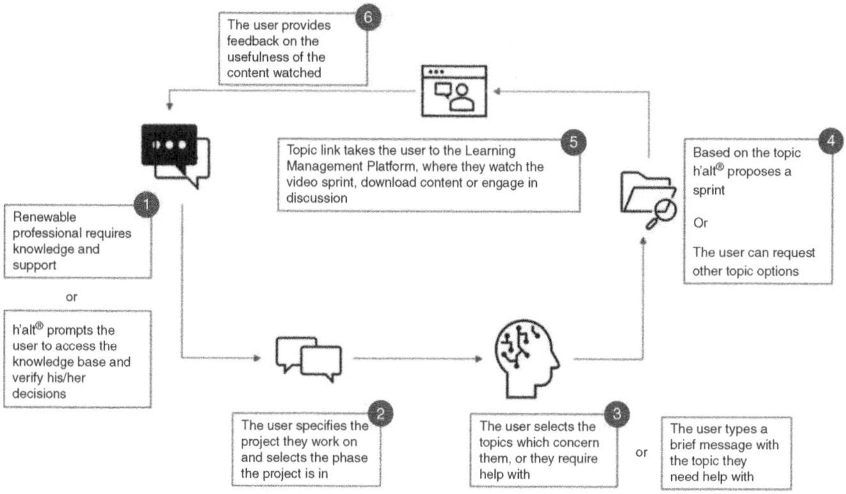

Figure 4.1 The user experience in h'alt®. *Source*: ARMSA Academy.

If they require support or knowledge to help them in making a specific decision or to confirm a decision they plan to take, they can type in a search question related to this. Or they may respond to a scheduled prompt. These "nudges" are scheduled during specific times to coincide with user planning cycles, such as the end of a week or the beginning of the next.

h'alt® prompts users to provide information about the work context, such as the project or task they are working on, location, asset type, and work phase. Responses to these contextualized search prompts provide key data to the organization, enabling patterns to emerge later in the analysis phase.

When users enter their questions, the system searches a bespoke content database. h'alt® then presents users with content that best matches the search term. Users watch short video content explaining the topic they asked to know more about.

After watching the sprint video, users are confident to proceed with their decision. Or they can speak to a subject matter expert within the organization if the content highlights anomalies in their original thought process or users recognize a material gap or shortfall.

Content that does not exist in the system is a strength, not a weakness. Unresolved requests produce a profile of high-frequency requirements, so the organization can decide what issues need addressing, what content needs to be created or if it is available elsewhere in the business and needs to be added to h'alt®, so that when people are searching for it, the content is then discoverable.

All the resolved and unresolved requests together also provide crucial reporting data, from which the organization can draw critical management insights, identifying key issues such as underperforming teams or managers, bottlenecks in processes, or safety concerns.

In summary, h'alt®:

- facilitates in-workflow learning via a machine learning-based mentor function that actively directs professionals to expert, role-based micro-content designed to inform, validate, or challenge key decisions in real time
- draws upon user familiarity to maximize engagement, so the technology accommodates a range of chat-based communications platforms, including Microsoft Teams
- analyses unresolved queries – identifying patterns and suggesting content to address these gaps, helping to improve the system's overall effectiveness
- drives organizational improvement by monitoring and assessing users' patterns of engagement, to identify knowledge gaps and other key risks – leveraging the patterns in big data to identify patterns to solve problems and produce actionable management insights.

Figure 4.2 shows how user engagement with the tool, as people ask it for support, helps to identify operational factors that are affecting performance at a local level. In multi-site, multi-jurisdictional renewable organizations, top-down approaches to management assume there are significant similarities across operations. In practice, there are more differences than similarities. These differences – in factors such as culture, contract structures, infrastructure, workforce readiness

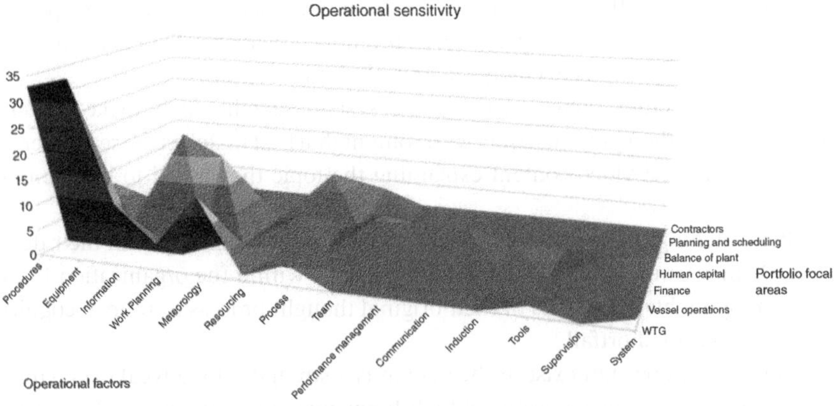

Figure 4.2 Aligning portfolio management with local needs across distributed renewable installations. *Source*: ARMSA Academy.

and availability, and supply chain maturity – all play a role in affecting local outcomes and determining the success of an asset.

h'alt® speeds up the ability of organizations to identify operational factors affecting work performance at a local level, not only ensuring that portfolio-level management decisions are directed toward supporting local needs but also rapidly developing greater operational sensitivity and awareness among middle and upper management.

4.2.4 Enabling Continuous Improvement Through ML and AI

h'alt® uses machine learning to continuously learn and improve based on user needs. It identifies gaps in knowledge, triggered by unresolved requests, and prioritizes content – learning from user interactions and improving its recommendations. h'alt® is using personalized (to the user's role and organization) knowledge and carefully curated content related to achieving safe outcomes in offshore wind, grounding the context that the AI will work with to the information retrieved from the source content and introducing automated evaluations to ensure truth and safeguards are maintained.

Although h'alt®, as an AI-enabled tool, is giving a clear understanding of where the gaps lie, human intervention is involved to ensure accuracy and relevance, and to do further analytics to identify and prioritize what gaps should be addressed. This is put back into the system, and people are learning from the additional content that's been put into the system. The level of knowledge and content within h'alt® is therefore growing according to the needs of the people who are using it.

There needs to be an element of human intervention to ensure that the gaps are valid gaps, and the content that is created to address those gaps must be curated accurately. The hazard profile of the industry is such that people's lives are at risk, and in safety-critical and risk-management scenarios, the last thing you want to do is inject more risk. The stochastic, rather than deterministic, nature of the environment means tools like h'alt® are still far away from relying solely on AI technology for complex decisions.

Generative AI can also assist with this type of tool; for example, it can help with content production. The original sprint content was produced as short-form videos using innovative lightboards and expert human tutors. Now the team is testing the use of software, such as Synthesia (www.synthesia.io), to create AI-generated video content using avatars and synthesized speech, which supports scalable, multilingual content delivery.

There is a wide range of avatar personas to choose from, so they could look like the existing industry training experts – a digital twin – or be tailored to the learning audience, so, for example, an Asian presenter for Asian markets. The use of AI in this context enables faster content creation and matching content to market, for example, translating content into different languages and making it more accessible. It also makes it quicker and simpler to adjust content to key changes, such as regulatory updates, and makes tools more adaptable to new geographic areas and markets.

Not only does AI generate the character presenting the content, but it also creates interchangeable backgrounds that match the words uploaded in the script. This generates videos that play behind the speaker. The quality of the production is high, and the technology has reduced content production time by around six to seven times. The increase in efficiency obviously must be balanced against the cost. But, as the technology is developing rapidly, it is becoming cheaper and more accessible. By removing pre- and post-production processes, content can be efficiently adjusted to suit other technologies such as utility-scale solar, battery storage, and others.

4.2.5 Next Steps

As decision-making tools develop, the potential to use AI more extensively is also evolving rapidly, with the caveat that using AI "just because we can" is not the objective. It is critical to make sure the application of AI enhances the system and contributes to the overall goals. As AI is developing so quickly, this requires extensive research into the technical aspects, but also considering costs and added value.

Overall, the key considerations going forward include:

- leveraging academic research and open-source advancements in AI to stay ahead of developments
- maintaining the balance between costs and performance
- emphasizing human-in-the-loop AI to maintain oversight and minimize risks.

In h'alt®, the case study tool, much of the work to derive the content and management insights from the patterns in user queries has been done manually. The next step is to bring more machine learning and AI into these types of platforms. Both the way the tool retrieves information when people are issuing queries and the way that the conversation is generated could be more AI-enabled, for example, using retrieval augmented generation (RAG) to improve the accuracy of AI responses by retrieving relevant data before generating answers, which ensures grounded and fact-based AI responses, crucial for user trust and transparency.

On the data reporting side, machine learning and AI can be used to better label the data and to identify insights beyond the structured data. As an example, an organization might have five categories of business goals to report against. The initial set of data is manually labeled by humans, and then the machine learns from those labeled datasets how to predict what labels should go against subsequent queries. Therefore, over time, it reduces the manual interaction required to label the data in preparation for reporting. The predictions the machines make have a probability of accuracy. So, by setting the probability threshold at, for example, 70%, this would make sure a prediction to label content of 70% or below would be flagged for human oversight, so this would continue to be human-in-the-loop AI.

After these fundamentals are done, that creates other opportunities. Once the reporting data is labeled, it allows more enriched visualizations of that data to give more perspectives of what is going on, making operational data particularly accessible to regional and global management teams because assigned labels relate to business goals, and key operational and functional levers such as equipment, resources, and contracts. It also means the AI can then reason over that data and assist in the analytics to suggest insights. This accelerates knowledge of where problems might be emerging and where there are knowledge gaps, giving organizations the opportunity to identify weaknesses and key leverage points earlier in the project lifecycle.

As the AI reviews the data, it is asked to find the patterns and correlations within the data and produce actionable insights. Again, this would be a first draft insight as part of the analytics. It would still require human oversight to make sure that it is correct and the machine is not hallucinating. It does, however, significantly speed up the process, and it might find some insights that a human may have overlooked. From a scaling perspective, this could be a game-changer for tools like this, because as the number of users and client organizations grows, having a human looking at every client's data would quickly become unsustainable.

As part of a proof-of-concept exercise, the team used Claude 3.5 Sonnet (www.anthropic.com/claude), by Anthropic, and uploaded anonymized data about unresolved queries together with workflow and activity context data. The large language model (LLM) was asked to examine the unresolved data and draw patterns of common observations, and to suggest content that might be uploaded to support those types of queries.

In total, there were 1638 searches, of which around 40% were satisfactorily resolved, which meant that there was content to address them. For around 800 or 900 searches, there was no content available. But the analysis identified 59 different topics, ranging from safety, systems, and processes to physical assets and technology issues that people need support on. Other practical examples raised include determining seasonal weather and operational mode-related themes, gaps arising in specific work types and locations, and assessing gaps in project planning tasks against governance frameworks. Without AI, this level of detail would not be possible.

In practice, if this was implemented within the tool using an LLM, it would be via an API (Application Programming Interface), a set of protocols for building and interacting with software applications, allowing different systems to communicate. This would include data privacy constraints.

Other options on the table are to explore using smaller models. For complex analytics, for example, an LLM might be appropriate. But a much smaller model could be fine-tuned specifically for categorizing. Partly due to the costs and environmental impact involved in using LLMs, people are realizing that not every task requires the biggest models. The most obvious and practical approach for tools like these is probably to use a mixture of fine-tuned small models and LLMs to get an acceptable balance between performance and cost.

Model selection is critical to the success of any AI solution. h'alt® is embracing an open-source AI approach and is using a pragmatic framework for addressing sovereign data concerns, agile change management, and the optimization of cost, performance, and energy consumption. Deployed via trusted cloud providers like Azure or through private infrastructures, this method ensures secure data control and robust compliance. Its flexible, streamlined architecture not only minimizes energy use but also circumvents vendor lock-in, allowing seamless integration with the h'alt® platform. This approach provides a sustainable and practical pathway for h'alt® to meet evolving AI demands effectively.

4.2.6 Results

After deploying h'alt® on two offshore wind operational projects – one service support team, and one special projects team – there were improvements in planning, collaboration, decision-making, and early identification of knowledge gaps.

Research from the Construction Industry Institute (2012) (see Table 4.1) shows that projects with high usage of front-end loaded decision-making, compared with those with low usage, achieved:

- 7% reduction of project delivery timelines,
- 10% cut in frictional costs, and
- 5% reduction in variations.

Table 4.1 Projects with high usage of front-end loaded decision-making vs low usage.

Efficiency factors	Reduction rate (%)
Project delivery timelines	7
Frictional costs	10
Variations	5

Adapted from the Construction Industry Institute (2012).

Findings from the deployment of h'alt® so far have broadly reflected this, showing reductions in project delivery times, frictional costs, and variations. There has also been a 60% decrease in management troubleshooting time.

The special projects team, comprising project managers, deputy project managers, and technical and non-technical staff, is responsible for planning and executing large-scale projects to maintain or upgrade turbine performance. Projects delivered by special projects include blade replacement campaigns, gearbox, transformer, generator exchange projects, and upgrades. For offshore locations, this requires hiring expensive jack-up barges to transport equipment from portside to affected turbines.

Insights provided by h'alt® have increased efficiency and safety in major component exchange campaigns, and reduced project schedules by helping to embed the special projects stage-gate model, which is the decision tree for these major component exchanges. After using the tool for 18 months, project managers have adopted the preferred stage-gate model to prepare, plan, and initiate projects. This sequential decision-making reduced a generator and major component exchange campaign at a Taiwanese offshore wind project by two days, saving c.400 000 Euro in jack-up barge (JUB) costs (based on average JUB daily hire rates). Other undisclosed project savings are attributed to a reduction in project headcount. No accidents or failures were reported during this project.

h'alt® has provided input to the offshore service support team, enabling the team to support service teams more effectively. Figure 4.3 illustrates the results after Sonnet 3.5 examined the unresolved data to draw patterns of common observations and show the relationships between key leverage points (points of intervention). This can support managers in identifying "keystones" to change, and suggest what effect acting on each keystone is likely to have on service and project operations. In practical terms, managers get insights into where real effort is required, which increases their ability to address problems effectively on the ground.

Feedback from h'alt® has helped to identify gaps in offshore training for technicians. The search queries revealed that technicians, who had passed formal technical training, were still asking fundamental technical questions that should have been part of their skillset. The response in this case was not to create new content for the tool, but to re-examine technician training to understand how they were trained and why these knowledge gaps still existed.

The data also identified an underperforming service manager, which meant it was possible to design a tailored performance management approach for the manager, who was based remotely and might not have been identified by other means. In hierarchical, high power distance cultures, people are less likely to complain formally or directly to line or senior managers. This highlights how the tool helps

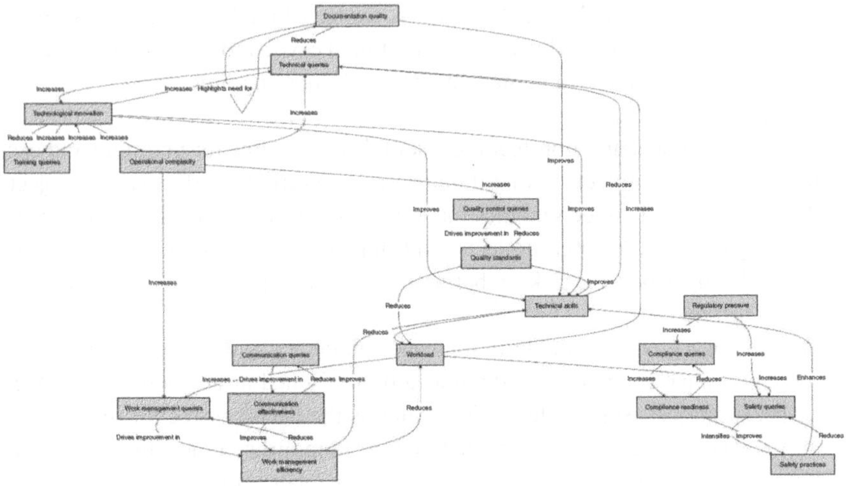

Figure 4.3 Causal loop diagram of high-level leverage points identified by tool searches (produced using Sonnet 3.5). *Source*: ARMSA Academy.

users to raise queries without being afraid of damaging team morale and managerial relationships. Via h'alt®, the problems came to light quickly, allowing the organization to identify the issue and rectify it early in the project.

Table 4.2 shows user responses following real-world implementation of the tool. Feedback from content consumers shows h'alt®:

- encourages early cross-functional collaboration and builds strong teams – 20% of users say the tool helps improve working relationships
- saves time – 22% of users say the tool helps them save time
- improves work quality – 30% of users say the content helps them improve quality.

4.2.7 After: What Is the Accepted Solution Now?

Digital decision-making tools are designed to support and influence people and relationships in complex adaptive organizations, so the changes that are happening are constantly evolving. There have already been changes within teams and networks, as well as shifts in relationships between roles, breaking down siloed thinking. One of the key observations is that h'alt® has affected how users (content consumers and senior management) behave, make decisions, and work with each other, building greater cohesion and collaboration.

Table 4.2 User feedback from real-world implementation.

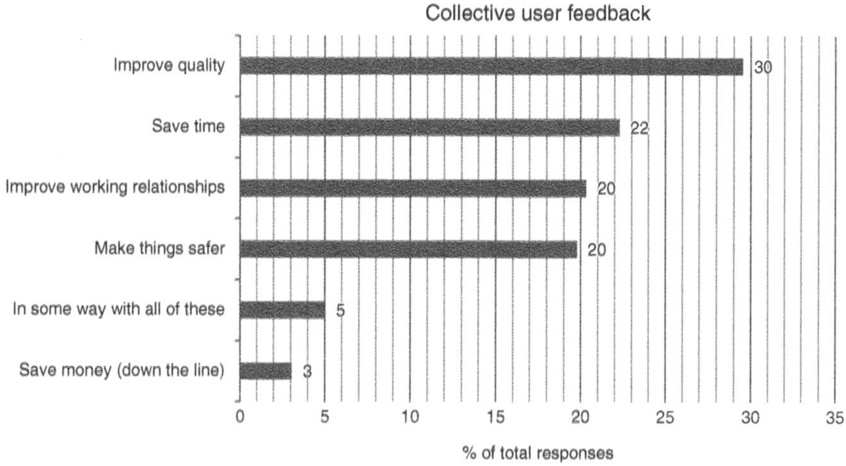

Source: ARMSA Academy
Note: Data correct as of August 2024

Over time, content consumers access the mentor chatbot more readily, suggesting they are valuing the content. Their search queries also become more well-formed and thought-out, as opposed to inputting one-word searches. This indicates they are thinking through the problem before making the request, which is helping them better structure the challenge they are facing and looking for support on.

Content consumers also gain confidence that senior managers examine and analyze their unresolved queries, on an anonymized basis. This means they start to use the tool to point managers more directly to problems that might be a bit more complex, and that require managers' involvement. This is particularly important in markets where a high power distance culture would dictate that people keep problems to themselves, either because they are afraid to question or challenge line managers, or because they want to protect the integrity of their managers.

Users can also build their own team networks around tools like these – encouraging team-based conversations and collaborative decision-making. For example, the special projects team set up monthly team "huddles" based around h'alt®. In these, managers provide the teams with feedback and resolve complex and non-straightforward cases. When teams are under pressure, this type of ongoing meeting that facilitates communication and cross-functional cooperation can strengthen cohesion and assist problem-solving.

Furthermore, teams look at the tacit value of the information they are generating from these "huddles" and more broadly at how, if they solve a problem together, this generates more content that needs to go into the tool for others to benefit from their queries and insights.

Organizations report:

- increased collaboration, consultation, and collective problem-solving help build an organizational bank of tacit knowledge
- accelerated team learning through quality collaboration rather than solely relying on individual experiences and mistakes
- decision-making now aligns with the complexity of their projects and business goals
- users constantly assess the effectiveness of processes, making them more adaptive and resilient
- feedback from queries alerts decision-makers when key social networks are under stress, ensuring issues are addressed before they impact performance.

4.3 Discussion

One of the keys to success in developing and implementing h'alt® has been industry collaboration. For these tools to be successful and effective, they need to be grounded in industry research, take a design-led approach, and be based on systems thinking.

In terms of developing the AI elements of the tool, being part of the open source community has been crucial to success. The ongoing work to incorporate AI effectively, safely, and ethically within tools like these is highly dependent on leveraging breakthrough research from AI researchers. When decisions are made to use techniques such as RAG, this must be based on solid research. And the open source community provides easy access to the latest research in AI from researchers around the world.

The need for industry collaboration comes back to ensuring that the expert content the tool is providing to users reflects what these users are experiencing in their working environments. A research-based user-centered approach is critical. To be successful, tools need to reflect users' practical, daily requirements and what is technologically and economically viable.

Another key point to remember is that decision-makers need to access independent, accurate, and role-specific information at critical points as they go about their daily tasks in the workplace. This is only possible if tools contain content focused on individual roles. So, before anything is produced, it is necessary to identify key roles within organizations.

To break down functional silos, it is also crucial to map out where decision points involve others. This allows different roles and departments to develop a mutual understanding and vocabulary to help them make cross-cutting and collaborative decisions. It is not just the individual who is being developed but the context of the organization within which they are being developed.

Ideally, tools should recognize and bring in an even wider context, connecting people across organizations and countries. Sharing the concerns, experiences, and challenges others have encountered in coming to critical decisions will further inform the process, allowing the user to stand back and take a much broader view.

In developing h'alt®, the first step was a gap analysis, which asked leading industry professionals about the challenges and aspirations they faced, how they developed decision-making capability to optimize performance and minimize risk, and what they looked for in terms of support.

The next step was to study key roles within the wind power market to identify individual and collective decisions that affect organizational effectiveness, plant efficiency, and operational safety. This was followed by pilot testing in a real-world application through collaboration with industry to ensure the content met user needs.

h'alt®'s development continues with implementation in close collaboration with industry as deployed in real-world renewables projects in different global locations. The nature of the environment and tool means it is crucial to understand organizational needs as they evolve. It is also important to continue refining and adapting processes in accordance with organizational and user requirements.

One of h'alt®'s strengths is its adaptability as a holistic solution to universal sectoral and organizational challenges. In its essence, the tool aims to minimize risk and optimize performance in rapidly growing businesses, so it is equally applicable across other renewables sectors – whether in generation, transmission, or distribution.

Like its application in offshore wind, the tool's content must be tailored to the specific industry and sourced directly from subject matter experts or the organization. The team behind h'alt®, which has expert knowledge across the energy sector, is already developing content for solar PV based on primary research data, secondary research, domain knowledge, and real-world experience.

As discussed in the case study, the hazard profile of the energy sector means that using automatically sourced or just LLM AI-generated content is not an option in this context. The content must be authentic, trusted, and authoritative, and sourced directly from industry experts. Where AI can help is when using trusted content to ground an LLM model, and also when the model is fine-tuned to define how it responds. This enables specialized, smaller models to help with automation of data labeling for enhanced reporting, query summaries to complement sprint content, and assisting in personalized knowledge validation after sprints are consumed (personalized quick quiz). Another way AI can help with transferability and scalability is with content delivery, for example, by increasing the speed and reducing the costs associated with the creation and production of industry-specific sprints and videos.

From the organization's point of view, another important contributor to success is that, before they start to implement these tools, they need to have a strong vision for, and agreement about, what they want to achieve over a given timeframe. Forward thinking and open-minded leadership is also key to success. These types of decision-making tools are likely to work well in organizations that embrace an intent-based leadership model. Developed most notably by Marquet (2012), intent-based leadership describes situations where leaders desire to make leaders of their followers, to give their followers agency, and where people across the organization engage and contribute to their full extent, and are confident, and given the means, to make decisions and take responsibility.

4.4 Conclusion and Future Work

If renewable businesses are to scale effectively, responsibly, and safely, they need to take a more proactive approach to operation, workforce planning, and skills development. And building and distributing organizational knowledge is vital for on-site decision-making in fast-paced, complex, and uncertain environments.

Digital decision-making tools that harness AI to enhance user experience and provide more granular data to generate actionable insights earlier in the project lifecycle can help organizations:

- scale effectively and safely globally, regardless of baseline maturity
- leverage user data to understand knowledge and skills gaps in dynamic and global environments at a local level
- reduce the gap between work-as-imagined, work-as-planned, and work-as-done.
- predict and prioritize risk precursors
- identify systemic levers requiring attention – allowing leadership to focus on the issues that matter
- create high-functioning teams across geographies
- foster cross-functional collaboration and cross-cultural cohesion

Decision-makers on the ground can:

- access engaging, tailored content in the workflow
- build stronger teams based on common knowledge and common vocabulary, globally

- better understand how their decisions create risk and opportunity
- make early-stage decisions based on front-end loaded decision-making
- feel listened to constructively and collectively
- develop personal knowledge by applying learning in practice

Recent G+ reports show an increase in hazard observation rates and accidents and incidents across the offshore wind sector. There has also been an increase in fatalities. Figure 4.4 explains why this might be happening. The decision-making context influencing operational control and safety outcomes is made early on in an organization's growth journey. Social culture, competitive strategy, contractual commitments, and obligations all play a direct role in how work is planned and executed.

It also shows that if operational impacts of workforce planning, contractual negotiations, organizational design, and work management are considered much earlier in the lifecycle, businesses would not be creating conditions where their operational demand outstrips their capacity. And even unforeseen emergent hazards and failure modes can be detected and mitigated well before work execution.

4.4.1 Challenges and Opportunities

Like the renewables sector, AI is developing at breakneck speed, so one of the key challenges ahead is to keep abreast of how AI technologies can practically, safely, and economically support or improve the way in which decision-making tools assist individuals, teams, and organizations.

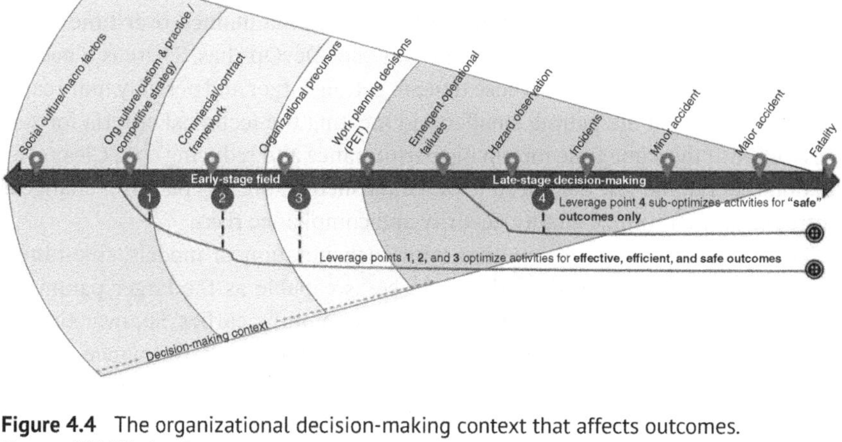

Figure 4.4 The organizational decision-making context that affects outcomes. *Source*: ARMSA Academy.

Currently, agentic AI is one of the most interesting developments on the horizon. This type of AI essentially involves a chain of separate, specialized AI agents that work together to complete complex tasks. Decision-making tools could potentially harness agentic AI to digitize the process further.

Within the tool, for example, there could be a series of agents, starting with an AI analytics agent carrying out the reporting data analysis and identifying content gaps. This could then communicate with a content creation AI tasked with researching and finding the various source materials. This content creation AI would then create a content plan and pass it on to another AI agent designed to write the content, which would then be reviewed by a content reviewer AI agent. In this way, agentic AI could offer significant advantages and efficiencies, as well as facilitate scaling. However, it also creates potential risks and requires robust coordination and cost-management strategies, as well as human oversight.

There are other ongoing AI challenges, which are broadly those facing all businesses adopting AI. These include:

- observability and evaluation, which are critical to understanding, monitoring, and analyzing applications, identifying problems, and optimizing quality and safety
- cost and performance – the high computational cost of using large AI models makes balancing performance and efficiency essential
- transparency and data privacy – data source transparency and compliance with privacy regulations, especially when using third-party APIs
- sustainability – the energy-intensive nature of AI models raises concerns, which are especially relevant in the renewable energy sector.

Dealing effectively and responsibly with continuous innovation is another clear challenge. The question about how AI models are maintained over time is still seeking answers to mature in the same way that DevOps has. There is a need to evaluate new models, not just those that are getting larger and possibly more capable, but those that are getting smaller and meeting the technical criteria for success while at the same time improving performance and reducing cost. Clear evaluation criteria and test data need to be established before any platform change is introduced to minimize quality, security and compliance risks.

The speed of change means that with each iteration of models, fine-tuned, smaller parameter models are shown to be as capable as the larger parameter models, for specific use cases, from just a few months earlier. So, over time, it should be possible to reduce the model size to maintain performance, reduce operational cost, and consume less energy. Recent advances in "inference time" improvements in model responses are also increasing the level of accuracy and capabilities of LLM models.

In terms of organizational and cultural barriers to adopting AI, fear of the technology can generally be assuaged if effective policies on AI are agreed upon at an organizational level. For example, if an organization agrees it will only use AI as a companion to assist in skills or analytics, but not as a direct displacement of humans, people tend to be willing to contribute to the knowledge that powers AI, rather than being defensive in sharing information.

A decision-making tool such as h'alt® is designed to support the human being and the organization on a learning journey. So, some of the barriers associated with other applications of AI are not present. One of the main sources of organizational or cultural risk aversion to AI is fear of hallucinations. But this type of decision-making tool does not use AI as a source of knowledge, but as a means of delivering that knowledge. So, again, that barrier is removed because the information provided to users is grounded in factual content that has come directly from industry experts and not from an LLM (where data used to train the model is often from web-scraped content).

The final challenge is not related to AI per se but to overcoming the wider potential barriers to organizations in adopting these tools. AI-enabled decision-making tools can be at the forefront of supporting businesses and their people as they scale at an unprecedented pace. But, so far, the renewables sector has been slow to recognize the benefits of investing in digital technologies that focus on human capital, rather than project assets.

AI that works is just software and, as such, falls under established security and governance policies that already exist in IT departments. Whether there is AI involved or not, organizations in the energy sector always have concerns about cybersecurity and breaches into systems. This is an ongoing issue for IT departments and can lead to a wider organizational reluctance to invest in innovative digital solutions.

Another barrier is the pressure in the renewables sector to deliver quick results. This makes some organizations unwilling to invest in new digital technologies that take up resources in the short term but are aimed at delivering medium to long-term benefits.

The barriers discussed above are not insurmountable but, to overcome them, industry leaders need to take an open-minded approach to improving organizational performance and resilience. Leaders need to have the desire to plan and execute now for the future. The first challenge is therefore a mindset challenge, and the second is the willingness and desire not to do something that has always been done before, but to choose something fundamentally different. Above all, the sector needs to recognize that its people are its greatest asset and that, as the skills gap widens, it cannot rely on traditional approaches to skills development and organizational improvement.

Acknowledgments

We thank Lucie Ponting for her contribution to this chapter.

Bibliography

ARMSA Academy. *Sprint Into the Future*, (2020). Available at https://armsa.academy/renewable-energy-skills-crisis/. (accessed 18 February 2025).

ARMSA Academy. *How can we develop high energy renewables projects*? (2024). Available at: https://armsa.academy/renewable-project-development-2024-research-reveals-critical-success-factors/. (accessed 14 November 2024).

J Bersin. *A New Paradigm for Corporate Training: Learning in the Flow of Work*, (2018). Available at: https://joshbersin.com/2018/Bersinb06/a-new-paradigm-for-corporate-training-learning-in-the-flow-of-work/. (accessed 14 November 2024).

J Bersin. Teams, not hierarchy. In *Irresistible: The Seven Secrets of the World's Most Enduring, Employee-Focused Organizations*, p. 9. Washington D.C., Ideapress Publishing, (2022a).

J Bersin. Teams, not hierarchy. In *Irresistible: The Seven Secrets of the World's Most Enduring, Employee-Focused Organizations*, p. 10. Washington D.C., Ideapress Publishing, (2022b).

Construction Industry Institute. *CII Best Practices Guide: Improving Project Performance*. United States of America, (2012). Available at: https://www.construction-institute.org/leading-industry-practices-for-estimating-controlling-and-managing-indirect-construction-costs-ve

G+ Global Offshore Wind Health and Safety Organisation. *2023 Incident Data Report*, (2024). Available at: www.gplusoffshorewind.com/?a=1538335. (accessed 14 November 2024).

IRENA. *World Energy Transitions Outlook 2023: 1.5 °C Pathway*. Abu Dhabi, International Renewable Energy Agency, (2023). Available at: www.irena.org/Publications/2023/Jun/World-Energy-Transitions-Outlook-2023. (accessed 14 November 2024).

LD Marquet. *Turn the Ship Around: A True Story of Turning Followers Into Leaders*. New York, Penguin Group, (2012).

P Senge. *The Fifth Discipline: The Art and Practice of the Learning Organization*, p. 3. New York, Doubleday/Currency, (1990a).

P Senge. *The Fifth Discipline: The Art and Practice of the Learning Organization*. In *Give Me a Lever Long Enough... and Single-Handed I Can Move the World*, p. 4. New York, Doubleday/Currency, (1990b).

www.anthropic.com/claude

www.synthesia.io

5

Relevance of AI in Addressing Barriers to Rooftop Solar Photovoltaic Adoption in Building Projects in Nigeria

Oluwadamilare Olamide Ilesanmi,[1] Ayodeji Emmanuel Oke,[2] Dorcas T. Moyanga,[3] John Aliu,[4] Lekan Damilola Ojo,[5] and Oladotun Nathanael Agboola[6]

[1] *Department of Civil Engineering, Faculty of Engineering, Lakehead University, Thunder Bay, Canada*
[2] *Department of Quantity Surveying, Research Group on Sustainable Infrastructure Management Plus (RG SIM +), Federal University of Technology Akure, Akure, Nigeria*
[3] *Department of Quantity Surveying, Federal University of Technology Akure, Akure, Nigeria*
[4] *Engineering Education Transformation Institute, College of Engineering, University of Georgia, Athens, GA, USA*
[5] *Department of Building and Real Estate, The Hong Kong Polytechnic University, Hong Kong, China*
[6] *Department of Construction and Real Estate Management, Metropolia University of Applied Sciences, Helsinki, Finland*
[1] *https://orcid.org/0009-0002-7916-8932*
[2] *https://orcid.org/0000-0001-6551-8634*
[3] *https://orcid.org/0000-0002-4351-4111*
[4] *https://orcid.org/0000-0001-5651-4009*
[5] *https://orcid.org/0000-0001-9154-8001*
[6] *https://orcid.org/0000-0002-5191-3294*

5.1 Introduction

The increasing challenges of climate change have prompted global efforts to adopt sustainable energy solutions (Usman et al., 2024). Among these, rooftop solar photovoltaic systems have emerged as a transformative technology, offering a clean, accessible, and cost-effective means of addressing the world's growing energy demands while mitigating the adverse impacts of fossil fuels (Wang and Liu, 2024). In both developed and developing countries, solar photovoltaic systems are being integrated into building projects as a pivotal component of the transition to renewable energy (Kumar et al., 2024; Akgayev and Unsur, 2024). While developed

Renewable Energy Transition with Artificial Intelligence: Challenge-driven Solutions, First Edition. Nina Dethlefs and Joyjit Chatterjee. Published 2026 by John Wiley & Sons, Inc.

countries have made significant progress in adopting rooftop solar photovoltaics, developing nations like Nigeria still grapple with numerous barriers to widespread implementation, particularly in building projects (Ohanu et al., 2024).

Despite abundant solar radiation and vast renewable energy potential, adopting rooftop solar photovoltaic systems remains limited in Nigeria (Ijeoma et al., 2023). Studies indicate that Nigeria receives an average solar radiation of 3.5–7.0 kWh/m^2 daily, making it one of the most solar-rich countries globally (Amole et al., 2023; Ohanu et al., 2024). However, the integration of this resource into building projects is hindered by a range of barriers, including high initial costs, lack of technical expertise, limited awareness, policy and regulatory challenges, and inadequate infrastructure (Roy and Pearce, 2024). These barriers not only impede the development of smart cities but also limit the potential of rooftop solar photovoltaics to address the country's energy deficit (Kumar et al., 2024).

Artificial intelligence (AI) has emerged as a game-changing tool with the potential to revolutionize the adoption of rooftop solar photovoltaics in Nigerian building projects (Tunde et al., 2024). AI applications can address key barriers by optimizing system designs, predicting costs, improving maintenance efficiency, and enhancing stakeholder awareness through intelligent solutions (Okpala and Nzeanorue, 2024). Consistent integration of rooftop solar photovoltaic systems across Nigerian building projects has the potential to generate over 600 000 MW of renewable energy annually, significantly reducing reliance on nonrenewable sources and fostering energy independence (Ideki et al., 2024). By leveraging AI, Nigeria can unlock the full potential of its solar resources, streamline adoption processes, and create a more sustainable energy landscape (Manuel et al., 2024).

However, despite the significant advancement of solar energy adoption worldwide, gaps remain in understanding how AI can be applied to address barriers in the Nigerian context. For example, studies in developed nations, such as the United States and the United Kingdom, have explored AI applications for optimizing solar energy systems and predicting performance (Aldarsi and Singh, 2024; Kayaçetin and Hozatlı, 2024). Similarly, research in Malaysia, South Africa, and China has highlighted the barriers to adoption and proposed strategies for overcoming them(Fauzi and Ismail, 2024; Mutumbi et al., 2024; Ren et al., 2024). In Nigeria, Manuel et al. (2024) explore the influence of AI on enhancing renewable energy utilization and visual power plant efficiency. Olaleye et al. (2024) evaluate the impact of roof tilt and building azimuth on the efficiency of off-grid photovoltaic power systems. Okpala and Nzeanorue (2024) assessed the implementation of IoT and AI for Sustainable Urban Development. However, no research categorizes these barriers into manageable segments and systematically maps AI solutions to each category. Furthermore, previous studies often overlook the comparative perspectives of construction professionals and users/clients, which are critical for practical implementation.

This study distinguishes itself by bridging these gaps. It categorizes barriers to rooftop solar photovoltaic adoption in Nigerian building projects and maps AI solutions to each identified barrier. Additionally, it compares the perspectives of construction professionals and clients, providing a balanced and inclusive framework for actionable solutions. The originality of this research lies in its dual focus on barriers and AI relevance, offering a comprehensive understanding of how technology can accelerate the adoption of rooftop solar photovoltaics in Nigeria. Ultimately, the study contributes to the discourse on renewable energy adoption, emphasizing the transformative role of AI in fostering sustainable building practices and advancing Nigeria's transition toward smart, energy-efficient communities.

5.2 Literature Review

5.2.1 Overview of Rooftop Solar Photovoltaic Systems

Based on the photovoltaic effect, rooftop solar photovoltaic systems convert sunlight into electricity (Amole et al., 2023). These systems typically consist of components such as solar panels, inverters, battery storage, charge controllers, utility meters, and connections to the grid (Roy and Pearce, 2024). There are three main system types – monocrystalline, polycrystalline, and thin-film – each designed to suit varying energy needs (Ibegbulam et al., 2023). Monocrystalline systems are highly efficient and space-saving, making them suitable for areas with limited roof space (Akhter et al., 2024). Polycrystalline systems, though slightly less efficient, are more cost-effective and ideal for budget-conscious projects (Li and Chen, 2024). Thin-film systems are lightweight and flexible, performing well in low-light conditions, but typically have lower efficiency compared to the other types (Naraharisetti et al., 2024). See Figure 5.1 for more details.

Solar PV systems are lauded for their environmental benefits, including zero-emission energy generation and reduced reliance on fossil fuels (Ijeoma et al., 2023). They also offer economic potential, from microgrid development to job creation in installation and maintenance (Usman et al., 2024). Despite their potential, the environmental impacts of production and end-of-life management of components remain concerns (Akgayev and Unsur, 2024). For developing countries like Nigeria, the abundant solar resource presents an opportunity to address energy poverty, enhance sustainability, and support economic growth (Ohanu et al., 2024).

5.2.2 Reluctance to Adopt Sustainable Energy Solutions in Nigeria

Nigeria's construction industry and communities exhibit a slow adoption rate for sustainable energy solutions like rooftop solar photovoltaics (Ijeoma et al., 2023). Limited awareness and inadequate education on the socioeconomic and

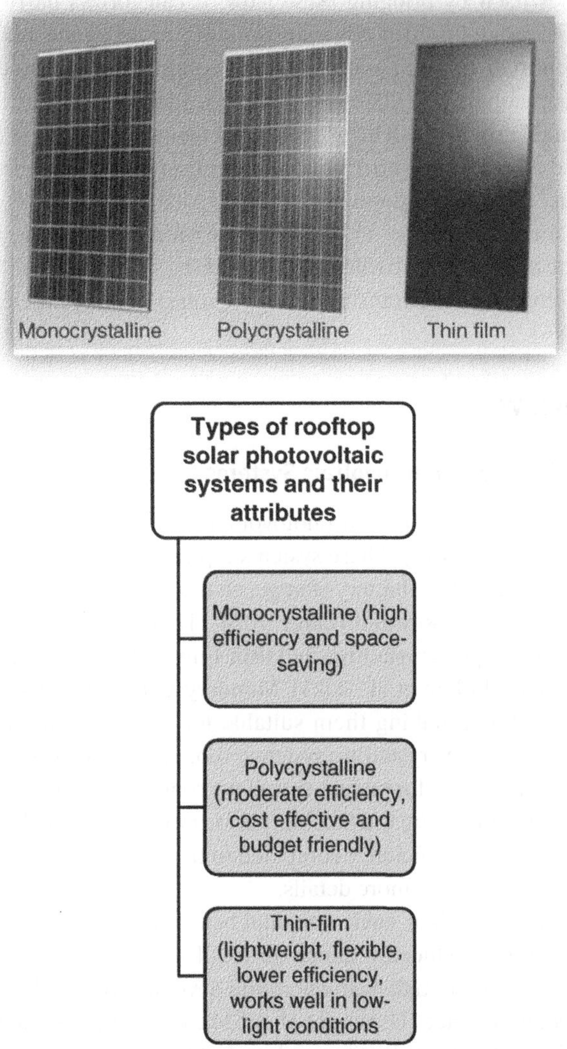

Figure 5.1 The three types of rooftop solar photovoltaic systems and their attributes.

environmental benefits of these systems are significant barriers (Olaleye et al., 2024). Mismanagement of existing solar installations due to substandard materials and unskilled labor has eroded public trust in the technology (Ohanu et al., 2024). Additionally, the high upfront costs of solar systems deter many potential adopters. Despite these challenges, solar photovoltaic adoption in Nigeria is gradually gaining traction (Manuel et al., 2024). However, the pace is slow, with adoption rates trailing behind many other developing nations. Factors such as the lack of government incentives, poor grid infrastructure, and insufficient financing models exacerbate the situation (Olaleye et al., 2024). Comparatively, countries like Kenya and Bangladesh have implemented creative financing mechanisms, such as pay-as-you-go systems, enabling broader access to solar PV technology (Ongeri and Mbataru, 2023; Hussain et al., 2024). For example, Kenya's pay-as-you-go model has reached over 500 000 households, increasing solar PV penetration to approximately 30% of rural communities (Rotich et al., 2024). Bangladesh has also achieved remarkable success, with over six million solar home systems installed under the IDCOL program, providing electricity to nearly 20% of its population (Hussain et al., 2024). In contrast, Nigeria's adoption rate remains below 1%, highlighting the stark difference in uptake and the missed opportunities for widespread solar deployment (Osuizugbo et al., 2024). Nigeria's reliance on conventional energy sources and limited policy frameworks to support renewable energy further hinder progress (Amole et al., 2023).

5.2.3 Barriers to Rooftop Solar Photovoltaic Adoption in Building Projects

The adoption of rooftop solar photovoltaic systems in Nigeria's building projects faces numerous challenges that limit their widespread integration (Olaleye et al., 2024). One significant barrier is the high initial cost of installation, which discourages many potential users and stakeholders (Mishrif and Khan, 2024). This is compounded by limited access to financing options, as financial institutions are often hesitant to provide loans for renewable energy projects due to perceived risks (Ijeoma et al., 2023). Poor grid infrastructure in Nigeria further hampers the integration of rooftop solar systems into existing energy networks, making them less attractive for building projects (Amole et al., 2023). Another barrier is the lack of skilled personnel to install and maintain solar photovoltaic systems, which leads to frequent technical failures and reduced efficiency (Ongeri and Mbataru, 2023). This issue is exacerbated by the prevalence of substandard solar components in the market, which undermines the reliability of the system's performance (Aldarsi and Singh, 2024). Inconsistent government policies and the absence of strong enforcement mechanisms for renewable energy adoption also deter stakeholders from investing in rooftop solar solutions (Ibegbulam et al., 2023). Cultural factors and low awareness about the benefits and reliability of rooftop solar systems present additional barriers (Manuel et al., 2024). Many stakeholders in building projects still prefer conventional energy sources, often due to misconceptions about the performance and cost-effectiveness of solar technology (Wang and Liu, 2024). Furthermore, environmental factors such as harsh weather conditions, including dust storms and heavy rainfall, can reduce the efficiency of solar panels (Usman et al., 2024). The lack of proper waste management systems for used or damaged solar components poses long-term environmental concerns (Roy and Pearce, 2024). Economic instability, including fluctuating exchange rates, increases the cost of importing solar components, while the absence of government subsidies or incentives further discourages adoption (Kumar et al., 2024). Limited integration of renewable energy policies into urban planning frameworks also creates challenges, as solar adoption is not adequately prioritized in building project designs (Okpala and Nzeanorue, 2024). These barriers highlight the need for innovative approaches to address the multifaceted challenges facing rooftop solar photovoltaic adoption in Nigeria.

5.2.4 Artificial Intelligence Solutions to Overcome Barriers to Rooftop Solar Photovoltaic Adoption

Artificial intelligence provides numerous solutions to address the challenges limiting rooftop solar photovoltaic adoption in Nigerian building projects. AI-powered predictive maintenance tools can monitor the performance of solar systems in real-time, identifying and resolving faults before they cause significant failures (Tunde et al., 2024). These tools enhance the reliability of solar installations, addressing concerns about frequent technical issues (Okpala and Nzeanorue, 2024). AI also optimizes energy storage by predicting energy demand and solar generation patterns, improving the integration of rooftop solar systems into existing energy networks (Ren et al., 2024). To address the high cost of installation, AI-driven financial models provide customized cost-benefit analyses for solar projects, demonstrating their long-term economic benefits (Manuel et al., 2024; Dada and Popoola, 2023). Automated credit risk assessments enable financial institutions to offer tailored loan options, making solar adoption more accessible to stakeholders (Mutumbi et al., 2024). AI applications in supply chain management reduce procurement costs by optimizing the sourcing and delivery of solar components, mitigating the impact of fluctuating exchange rates (Van Tran et al., 2024). AI-based educational platforms play a critical role in increasing awareness about rooftop solar systems (Sheng and Liu, 2023; Chen et al., 2023). These platforms deliver targeted information to various stakeholders, addressing misconceptions and building trust in technology (Kumar et al., 2024). Sentiment analysis tools, powered by AI, identify public perceptions and concerns, enabling more effective communication and awareness campaigns. Virtual reality tools supported by AI allow users to visualize how solar systems would look and function in their buildings, encouraging acceptance and adoption (Okpala and Nzeanorue, 2024). Inconsistent policies and the lack of integration into urban planning frameworks can be mitigated through AI-driven urban energy planning tools (Shen et al., 2024; Mustafa et al., 2024). These tools analyze data to recommend optimal locations and designs for rooftop solar systems, ensuring their alignment with broader energy policies

(Van Tran et al., 2024). AI also supports policymakers by simulating the impact of different policy scenarios, helping to develop effective and enforceable renewable energy standards. Environmental challenges, such as the impact of harsh weather on solar panel efficiency, can be addressed with AI algorithms that forecast weather conditions and adjust energy production accordingly (Thakur and Wilson, 2024). AI also contributes to developing waste management strategies for solar components, promoting environmental sustainability (Freitas et al., 2023). By leveraging these AI solutions, Nigeria can overcome the barriers to rooftop solar photovoltaic adoption, paving the way for more sustainable building projects.

The specific barriers hindering the adoption of rooftop solar photovoltaic systems in Nigeria's construction industry and communities was highlighted in Table 5.1. It also provides AI-based solutions tailored to address each barrier effectively, as supported by relevant literature sources. The barriers encompass technical, economic, social, and policy-related challenges, while the proposed AI solutions range from virtual training simulations to market modeling tools and recycling optimization algorithms, demonstrating the multifaceted role of AI in overcoming these challenges.

Table 5.1 Comprehensive overview of barriers to rooftop solar photovoltaic adoption in Nigeria and AI-preffered solutions.

S/N	Barriers	AI-preferred solution	Literature sources
1	Unskilled labor	Virtual training simulations	Ongeri and Mbataru (2023); Okpala and Nzeanorue (2024)
2	Grid infrastructure issues	Smart grid integration	Amole et al. (2023); Manuel et al. (2024)
3	Lack of awareness and education	AI-driven educational platforms	Roy and Pearce (2024); Sheng and Liu (2023)
4	Corruption	Transparent blockchain tracking system	Manuel et al. (2024); Pal et al. (2023)
5	Intermittency of solar radiation	Weather-based predictive energy systems	Fauzi and Ismail (2024); Van Tran et al. (2024)
6	Operation and maintenance cost	Predictive maintenance analytics	Kumar et al. (2024); Ren et al. (2024)
7	Limited financing options	Personalized loan recommendations	Ijeoma et al. (2023); Mutumbi et al. (2024)
8	Government policy	Data-driven policy recommendation	Ibegbulam et al. (2023); Shen et al. (2024)

(*Continued*)

Table 5.1 (Continued)

S/N	Barriers	AI-preferred solution	Literature sources
9	Supply chain constraints	Demand forecasting tools	Pal et al. (2023); Kumar et al. (2024)
10	Social perceptions	AI-powered community engagement campaign	Aldarsi and Singh (2024); Ijeoma et al. (2023)
11	High initial cost	Dynamic pricing and financing models	Mishrif and Khan (2024); Tunde et al. (2024)
12	Economic instability	AI-assisted market analysis	Tunde et al. (2024); Van Tran et al. (2024)
13	Policy inconsistencies	Sentiment analysis for stakeholder input	Wang and Liu (2024); Thakur and Wilson (2024)
14	Environmental concerns	Recycling optimization algorithms	Hussain et al. (2024); Freitas et al. (2023)
15	Limited research and development	AI-augmented innovation platforms	Ibegbulam et al. (2023); Chen et al. (2023)
16	Cultural barriers	AI-behavioral insights	Manuel et al. (2024); Fauzi and Ismail (2024)
17	Energy Pricing Policies	Market modeling and simulation tools	Okpala and Nzeanorue (2024); Thakur and Wilson (2024)
18	Lack of collaboration	Partnership mapping and network analysis	Mishrif and Khan (2024); Pal et al. (2023)
19	Limited access to technology	AI-powered distribution planning	Olaleye et al. (2024); Manuel et al. (2024)
20	High import costs	Cost optimization through AI prediction	Mishrif and Khan (2024); Dada and Popoola (2023)
21	Weather dependence	Seasonal energy storage recommendation	Usman et al. (2024); Thakur and Wilson (2024)
22	Land use issues	Rooftop suitability assessment	Mutumbi et al. (2024); Mustafa et al. (2024)

5.3 Research Methods

This study addresses the barriers to adopting rooftop solar photovoltaic systems in Nigerian building projects and explores the relevance of artificial intelligence in overcoming these challenges. It adopts a quantitative research strategy using a structured questionnaire to collect data from construction professionals and clients in Lagos and Port Harcourt. These cities were chosen for their unique characteristics: Lagos serves as Nigeria's commercial hub with notable advancements in smart infrastructure (Xiao, 2021), while Port Harcourt is known for its growing focus on sustainable and energy-efficient buildings (Onwuadiochi et al., 2021). By targeting these dynamic contexts, the study provides insights into the challenges of adoption and the role artificial intelligence could play in addressing them.

The questionnaire was designed to capture responses about barriers and the perceived relevance of artificial intelligence. A 5-point Likert scale was employed for consistency and clarity. For barriers, the scale ranged from 5 ("strongly agree") to 1 ("strongly disagree"), while for the relevance of artificial intelligence, the scale ranged from 5 ("highly relevant") to 1 ("not relevant at all"). This approach ensured that responses reflected nuanced perceptions of both the challenges and potential solutions (Alabi and Jelili, 2023). The questionnaire was distributed via Google Forms, achieving a 73% response rate, with 155 responses retrieved out of 213 distributed. This response rate was satisfactory for the study's objectives.

Python was used extensively for data analysis due to its versatility and efficiency in handling large datasets (May et al., 2022). Descriptive statistics, such as percentages and frequencies, were calculated using the Pandas library to summarize the responses. The ranking of barriers was performed using the mean item score (MIS) and standard deviation (SD), which provided a straightforward way to identify the most significant challenges based on respondents' perceptions. Visualization of results, such as bar charts for ranked barriers, was achieved using Matplotlib and Seaborn libraries. Exploratory factor analysis (EFA) was conducted to group the 22 identified barriers into clusters, helping to simplify and interpret the data. This analysis was performed using Python's FactorAnalyzer package, which allowed for precise factor extraction and rotation, ensuring meaningful categorization of barriers. The clusters revealed patterns that would have been challenging to identify through simple descriptive methods, providing deeper insights into the nature of the barriers. To examine differences between the two respondent groups construction professionals and clients, the Mann–Whitney U test was utilized. This nonparametric test, implemented using Python's SciPy library, is particularly suited for ordinal data and does not require assumptions about normal distribution. The test assessed whether perceptions of barriers and the relevance of artificial intelligence differed significantly between these two groups (Raschka et al., 2020).

By employing this method, the study provided a robust analysis of group-specific variations, enhancing the depth of interpretation. In summary, Python's capabilities were integral to the study's analysis as indicated in Figure 5.2. Its tools for descriptive statistics, factor analysis, and hypothesis testing ensured accurate and

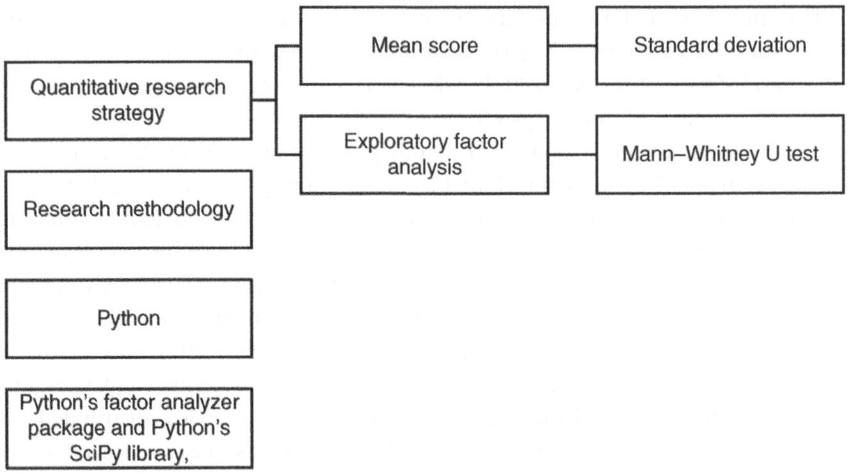

Figure 5.2 Research design framework and analytical framework.

efficient data handling. The methodology, grounded in clear objectives and consistent analytical approaches, provides a solid foundation for understanding the barriers to rooftop solar photovoltaic adoption and the relevance of artificial intelligence in overcoming them.

5.4 Results and Findings

5.4.1 Background of the Respondents

An overview of the respondents' professional backgrounds and years of experience was provided in Figure 5.3. The respondents are divided into two main groups: professionals (quantity surveyors, architects, builders, and engineers) and clients. The professionals collectively make up 51.0% of the respondents, while the clients account for 49.0%. Among the professionals, engineers formed the largest subgroup (19.4%), followed by architects (12.9%), quantity surveyors (10.97%), and builders (7.74%). In terms of years of experience, as displayed in Figure 5.4, most respondents (36.8%) had 1–5 years of experience, while 28.4% had 6–10 years. Those with 11–15 years accounted for 23.2%, and only 11.6% had over 15 years of experience. On average, the respondents had approximately 10.5 years of professional experience. This distribution highlights a balanced representation between professionals and clients, offering diverse perspectives for the study.

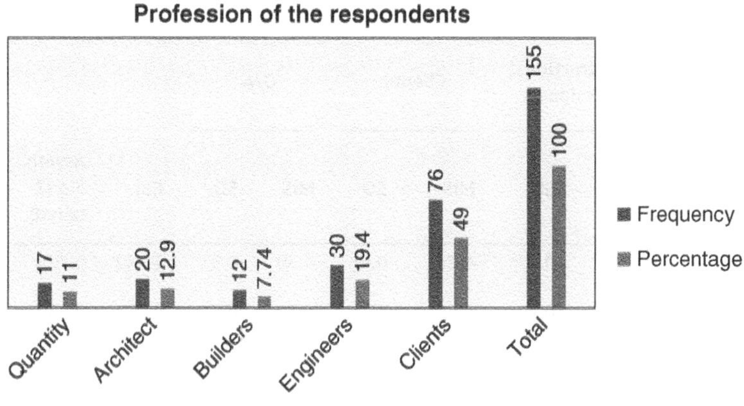

Figure 5.3 Distribution of respondents by profession (construction professionals and clients).

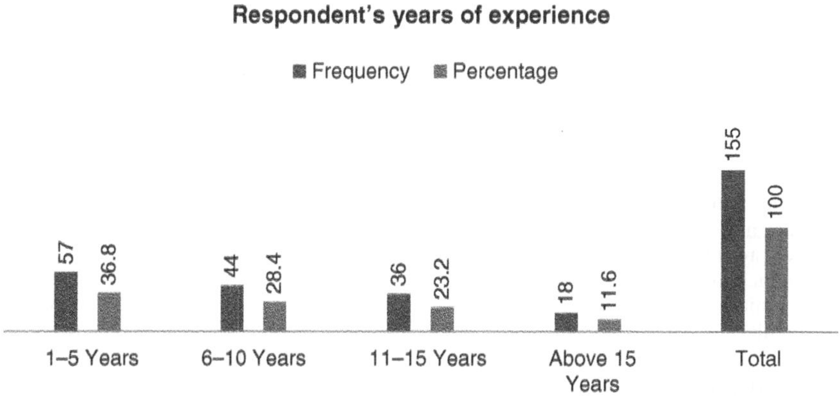

Figure 5.4 Distribution of respondents by years of experience.

5.4.2 Background Information of the Respondents

5.4.3 Barriers to the Adoption of Rooftop Solar Photovoltaics and the Preferred AI-Solution

The analysis of Table 5.2 highlights the barriers to adopting rooftop solar photovoltaics based on MIS, SD, and the Mann–Whitney U test.

5.4.4 Mean Score (MIS) Analysis

The top five barriers to adoption, ranked by their overall mean importance scores (MIS), are high initial cost (MIS: 4.30), limited financing options (MIS: 4.14), poor grid infrastructure (MIS: 4.10), lack of skilled personnel (MIS: 4.03), and

Table 5.2 Barrier to the adoption of rooftop solar photovoltaics.

	Construction professionals		Clients		O/A			
Variables	**MIS**	**SD**	**MIS**	**SD**	**MIS**	**SD**	**Chi**	**Asymp. Sig (2-tailed).**
The high initial cost	4.33	0.37	4.27	0.98	4.30	0.93	1.342	0.765
Limited financing options	4.21	0.89	4.06	0.76	4.14	0.65	0.324	0.059
Poor grid infrastructure	4.19	1.13	4.01	0.26	4.10	0.48	0.619	0.025*
Lack of skilled personnel	4.06	0.56	3.99	0.99	4.03	0.79	0.896	0.115
Government policies	4.01	0.87	3.88	0.88	3.95	1.10	0.007	0.884
Operation and maintenance costs	3.94	0.47	3.87	0.91	3.91	1.12	1.221	0.094
Intermittency of solar radiation	3.82	1.14	3.76	0.54	3.79	0.65	0.765	0.378
Social perception	3.72	0.94	3.65	1.01	3.69	0.66	1.922	0.456
Supply chain constraints	3.64	1.26	3.54	0.72	3.59	1.17	1.324	0.093
Lack of awareness and education	3.61	0.41	3.50	0.76	3.56	0.69	0.057	0.524
Economic instability	3.60	0.99	3.44	0.79	3.52	0.91	0.773	0.245
Policy inconsistencies	3.56	1.07	3.45	1.03	3.51	0.34	1.442	0.444
Environmental concerns	3.52	0.93	3.41	0.89	3.47	0.69	1.025	0.001*
Limited research and development	3.48	0.65	3.40	1.08	3.44	1.06	1.043	0.789
Cultural barriers	3.45	0.73	3.32	1.12	3.39	0.55	1.335	0.567
Energy pricing policies	3.41	1.22	3.30	0.43	3.36	0.83	2.442	0.345
Lack of collaboration	3.38	1.43	3.21	0.80	3.30	0.97	0.768	0.789
Limited access to technology	3.64	0.67	2.67	0.14	3.16	0.65	0.567	0.018*

	Construction professionals		Clients		O/A			
Variables	**MIS**	**SD**	**MIS**	**SD**	**MIS**	**SD**	**Chi**	**Asymp. Sig (2-tailed).**
High import costs	3.59	0.85	2.56	0.35	3.08	0.58	1.567	0.068
Weather dependence	3.46	0.64	2.45	1.15	2.96	0.79	0.679	0.714
Corruption	3.17	0.83	2.24	0.52	2.71	0.12	0.784	0.459
Land use issues	3.09	1.24	2.12	0.89	2.61	1.49	0.233	0.234

Note: O/A – Overall; * – $p < 0.05$.

government policies (MIS: 3.95) as indicated in Table 5.2. These barriers reflect critical issues in financing, technical capacity, and policy frameworks. The high initial cost is considered the most significant, reflecting the financial burden faced by users. Limited financing options exacerbate this challenge, making it difficult for stakeholders to access funding. Poor grid infrastructure impedes the integration of solar systems, while the lack of skilled personnel highlights the need for technical expertise. Unfavorable government policies further complicate adoption efforts. Artificial intelligence offers potential solutions to these barriers, including cost optimization and predictive analytics for affordable designs, AI-enabled credit scoring, and financial modeling to enhance access to financing, real-time grid monitoring systems to address infrastructure issues, AI-powered training platforms to develop technical skills, and policy modeling tools to optimize regulations.

5.4.5 Standard Deviation (S.D.) Analysis

As displayed in Table 5.2, the standard deviation values indicate the level of agreement among respondents. Barriers such as high initial cost (SD: 0.93) and limited financing options (SD: 0.65) have relatively low SD values, showing strong consensus among professionals and clients. However, some barriers exhibit higher variability, such as poor grid infrastructure (SD: 1.13 for professionals) and social perception (SD: 1.01 for clients), suggesting diverse opinions within the respondent groups. This variation may stem from differences in professional experience, access to resources, or personal circumstances.

5.4.6 Mann–Whitney Test Analysis

The Mann–Whitney U test compares responses between construction professionals and clients, revealing significant differences in three barriers: poor grid infrastructure ($p = 0.025$), environmental concerns ($p = 0.001$), and limited access to technology ($p = 0.018$), as shown in Table 5.2. These differences highlight the varying perspectives of the two groups, with professionals likely perceiving grid infrastructure issues as more critical due to their technical responsibilities, while clients are more sensitive to environmental concerns. The disparity in perceptions of technology access reflects structural inequalities that must be addressed to promote equitable adoption. The Mann–Whitney U test was conducted at a 0.05 level of significance ($p < 0.05$), meaning that any p-value less than 0.05 indicates a statistically significant difference between the two independent groups (Ilesanmi et al., 2024).

5.4.7 Exploratory Factor Analysis

To address the barriers to the adoption of rooftop solar photovoltaics in Nigeria, exploratory factor analysis (EFA) was employed to refine and categorize these barriers. The aim was to uncover hidden patterns and relationships, which could ultimately improve the implementation of rooftop solar technologies in Nigerian building projects. A Kaiser–Meyer–Olkin (KMO) value of 0.865, as shown in Table 5.3, indicated that the sample was adequate for factor analysis, while Bartlett's test results confirmed the data's suitability. The data's reliability was further verified with Cronbach's alpha score of 0.844, ensuring that the dataset was appropriate for this analysis. After meeting the required criteria, the factor analysis was carried out using principal component analysis (PCA) with varimax rotation, a method widely used for identifying underlying structures by detecting relationships between variables. The PCA revealed four main components, each with eigenvalues greater than 1. These four components accounted for a total of 71.23% of the variance, surpassing the 50% threshold recommended by (Stern et al., (2012); Ilesanmi et al., (2025). These four components were grouped into the following clusters, as seen in Figure 5.5:

Cluster 1: Technical Barriers
Cluster 2: Economic Barriers
Cluster 3: Social and Cultural Barriers
Cluster 4: Policy and Environmental Barriers

Table 5.3 Exploratory factor analysis.

KMO measure of sampling adequacy			**0.865**		
Bartlett's test of sphericity	Approximate chi-square		4821.654		
	Df		327		
	Sig		0.000		
Barriers					
	1	2	3	4	
Cluster1: Technical Barriers	–	–	–	–	
Grid infrastructure issues	0.867	–	–	–	
Intermittency of solar radiation	0.837	–	–	–	
Operation and maintenance costs	0.784	–	–	–	
Limited access to technology	0.768	–	–	–	
Limited research and development	0.732	–	–	–	
Weather dependence	0.719	–	–	–	
	–	–	–	–	
Cluster 2: Economic Barriers	–	–	–	–	
High initial costs	–	0.788	–	–	
Limited financing options	–	0.767	–	–	
Economic Instability	–	0.732	–	–	
Energy pricing policies		0.729			
Supply chain constraints	–	0.711	–	–	
High import cost	–				
	–	–	–	–	
Cluster 3: Social and Cultural Barriers	–	–	–	–	
Lack of awareness and education	–	–	0.778	–	
Unskilled labor	–	–	0.766	–	
Social perceptions	–	–	0.746	–	
Cultural barriers	–	–	0.741	–	
Lack of collaboration	–	–	0.728	–	
	–	–	–	–	
Cluster4: Policy and Environmental Barriers	–	–	–	–	
Government policy	–	–	–	0.761	
Policy inconsistencies	–	–	–	0.744	
Land use issues	–	–	–	0.728	
Environmental issues	–	–	–	0.713	
Corruption	–	–	–	0.705	

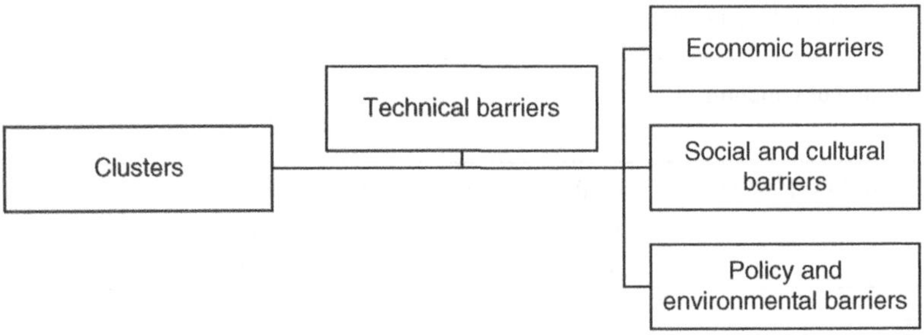

Figure 5.5 Clusters of barriers to rooftop solar photovoltaic adoption derived from the exploratory factor analysis.

5.4.8 Discussion and Implications of Findings

This study highlights barriers to rooftop solar adoption in building projects in Nigeria, explores artificial intelligence solutions, and provides ideas for policymakers and stakeholders to draw comparisons with global regions to inform sustainable strategies.

5.4.9 Mean Score

The mean score analysis ranked the barriers to rooftop solar photovoltaic adoption in Nigerian building projects. This revealed the five most critical barriers: high initial cost, limited financing options, poor grid infrastructure, lack of skilled personnel, and government policies. High initial costs were the most significant, reflecting the financial burden faced by stakeholders (Tunde et al., 2024). Limited financing options further exacerbated the challenge of affordability, while unskilled labor and lack of awareness pointed to gaps in technical capacity and education (Olaleye et al., 2024). Grid infrastructure issues underscored challenges with integrating solar photovoltaics into Nigeria's energy system (Manuel et al., 2024). Comparing these findings with other regions reveals differences and similarities. In South Asia, such as India, high initial costs and limited financing options are critical barriers, but government subsidies and innovative microfinancing schemes have somewhat mitigated these (Sun et al., 2021). In Latin America, particularly in Brazil, social barriers such as cultural resistance are more prominent than financial challenges (Mogrovejo-Narvaez et al., 2024). In contrast, other regions like Germany and Spain emphasize environmental concerns and grid integration efficiency due to their advanced financing and skilled labor systems (Silveira et al., 2024). Sub-Saharan African countries like Kenya face similar barriers to Nigeria, with high costs and low awareness dominating, but limited government interventions compared to Nigeria (Ongeri and Mbataru, 2023). However, policymakers and construction stakeholders in Nigeria, leveraging artificial intelligence tools like dynamic pricing and financing models, personalized loan models, smart grid integration, virtual training simulations, and data-driven policy recommendations, can address government's inconsistent policy and enhance rooftop solar photovoltaic adoption.

5.4.10 Exploratory Factor Analysis

Factor analysis grouped the 22 barriers into four clusters: technical barriers, economic barriers, social and cultural barriers, and policy and environmental barriers, explaining 71.23% of the total variance. Each cluster highlights critical challenges and reveals underlying patterns affecting rooftop solar photovoltaic adoption in Nigerian building projects.

Technical Barriers: This cluster includes issues such as grid infrastructure challenges, operation and maintenance costs, limited access to technology, limited research and development, weather dependence, and intermittency of solar radiation. These barriers highlight the technical limitations of Nigeria's current energy system, making integration of solar photovoltaics into the grid difficult (Okpala and Nzeanorue, 2024). For instance, unpredictability in solar energy generation due to weather conditions increases operational risks (Usman et al., 2024). In Europe, technical barriers are less significant due to advancements in grid technology and storage solutions, which allow seamless integration of solar photovoltaics (Nyffenegger et al., 2024). In contrast, in sub-Saharan Africa, such as Ghana, technical challenges are similar to Nigeria's due to weak infrastructure, but Ghana benefits slightly from regional collaborations to enhance grid development (Dokyi and Sharifi, 2024).

Economic Barriers: This cluster includes high initial costs, energy pricing policies, supply chain constraints, economic instability, limited financing options, and high import costs, which significantly hinder affordability. The lack of affordable financing models worsens the adoption rate, as many stakeholders cannot afford the upfront investment (Amole et al., 2023; Oke et al., 2024). In countries like Malaysia, economic barriers are mitigated through government subsidies and innovative financing schemes, reducing the impact of high initial costs (Fauzi and Ismail, 2024). In contrast, China's economic barriers are less severe due to a well-developed financing system, but fluctuations in energy pricing pose different economic challenges (Ren et al., 2024).

Social and Cultural Barriers: This cluster includes lack of awareness and education, lack of collaboration, social perception, unskilled labor, and cultural resistance. Misinformation about the benefits of rooftop solar photovoltaics and societal reluctance to adopt new technologies are significant issues in Nigeria

(Okpala and Nzeanorue, 2024). In the United States, cultural resistance is also prominent, as communities often mistrust new technologies, similar to Nigeria (Aldarsi and Singh, 2024). In East Asia, such as Japan, social and cultural barriers are minimal due to widespread public awareness campaigns and educational initiatives promoting solar energy (Numata et al., 2020).

Policy and Environmental Barriers: This cluster includes policy inconsistencies, corruption, and environmental concerns, such as recycling inefficiencies and land use conflicts. Inconsistent policies create uncertainty for investors, while environmental barriers deter sustainable practices. Policy inconsistencies and corruption in South Africa also dominate as barriers, mirroring Nigeria's challenges (Mutumbi et al., 2024). In contrast, Scandinavian countries like Denmark, Sweden, and Norway have clear and consistent policies to promote renewable energy adoption, with minimal environmental barriers due to strict regulations and recycling programs (Formolli et al., 2021).

Thus, these findings demonstrate the need for targeted solutions. Policymakers and stakeholders in Nigeria's construction industry can leverage artificial intelligence to address these clusters. For instance, predictive analytics can tackle technical issues, while AI-driven policy recommendations can streamline regulations and reduce environmental impacts. Collaborative efforts can enhance rooftop solar adoption and foster sustainable building practices.

5.4.11 Mann–Whitney U Test

The Mann–Whitney U test examined significant differences in perceptions between construction professionals and clients regarding variables such as poor grid infrastructure, environmental concern, and limited access to technology. These findings highlight diverse stakeholders' perspectives on adoption barriers. Comparative studies in other regions reveal unique patterns. For instance, in Jordan, clients are more concerned with affordability, while professionals prioritize grid reliability (Mashal et al., 2023). In India, professionals and clients diverge on environmental priorities, with clients emphasizing sustainability over cost (Parsad et al., 2020). Both groups share similar concerns over economic and technical barriers in African nations like Ghana, aligning closely with Nigeria's findings (Dokyi and Sharifi, 2024). Ultimately, the alignment between professionals and clients in Nigeria presents an opportunity for unified decision-making. Artificial intelligence tools such as sentiment analysis and stakeholder network mapping can bridge gaps, foster collaboration, and address barriers effectively, improving the adoption of rooftop solar photovoltaics in Nigerian building projects.

5.4.12 Harnessing the Power of AI to Overcome Barriers to Rooftop Solar Photovoltaic Adoption

Artificial intelligence holds immense potential to address barriers to rooftop solar photovoltaic adoption, especially in developing regions like Nigeria. AI can enable real-time data analysis for optimizing energy output and identifying technical inefficiencies. It can also facilitate predictive maintenance by identifying faults in rooftop solar systems before they escalate. Moreover, AI-powered platforms can provide tailored financial models, offering users insights into cost-effective options based on their energy consumption. AI-driven social media campaigns and recommendation systems can reshape public perceptions, countering cultural biases and misinformation about solar energy. These tools can help educate communities, demonstrate benefits, and drive widespread adoption. By integrating AI into decision-making processes, stakeholders can create efficient, inclusive, and sustainable frameworks for renewable energy implementation in the global south (see Figure 5.6).

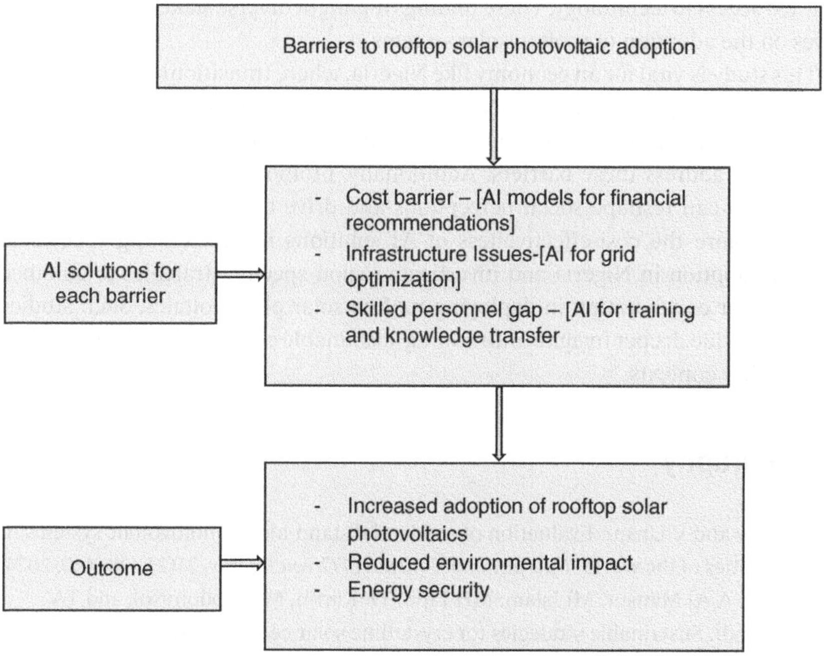

Figure 5.6 The role of artificial intelligence in overcoming barriers to rooftop solar photovoltaics adoption in Nigeria.

5.5 Conclusion and Perspectives

Renewable energy, such as rooftop solar photovoltaics, addresses energy shortages and environmental challenges in developing countries like Nigeria. Leveraging artificial intelligence offers innovative solutions to overcome barriers to adoption, making it a vital tool in fostering sustainable energy systems. This study aimed to identify barriers to rooftop solar adoption in Nigeria and explore AI-driven solutions to address them, utilizing insights from 22 barriers and corresponding AI-preferred solutions drawn from recent literature. The study employed Python for data analysis and used tools such as Mean Item Score Analysis, EFA, and Mann–Whitney U test.

The key findings revealed that high initial cost, limited financing options, poor grid infrastructure, lack of skilled personnel, and government policies were the top barriers to adoption. EFA grouped the barriers into four clusters: technical and economic barriers, social and cultural barriers, and policy and environmental barriers, which explained 71.23% of the total variance. The Mann–Whitney U test showed significant differences between construction professionals and clients regarding poor grid infrastructure, social perception, environmental concerns, and limited access to technology. These findings highlight diverse stakeholder perspectives on the adoption of rooftop solar systems.

This study is vital for an economy like Nigeria, where transitioning to renewable energy can mitigate energy challenges and reduce environmental impact. Policymakers should prioritize investments in smart grid technologies and supportive policies to address these barriers. Additionally, promoting AI-driven awareness campaigns can reshape social perceptions and drive adoption. Future research could explore the cost-effectiveness of AI solutions in accelerating renewable energy adoption in Nigeria and investigate region-specific strategies to enhance stakeholder collaboration in deploying rooftop solar photovoltaics. Such studies would provide deeper insights into scaling sustainable energy solutions in Nigeria and similar contexts.

Bibliography

B Akgayev and V Unsur. Evaluation of residential stand-alone photovoltaic systems in mega cities of the world. *International Journal of Green Energy*, 21(2):448–460, 2024.

M Akhter, A Al Mansur, MI Islam, MH Lipu, TF Karim, MG Abdolrasol, and TA Alghamdi. Sustainable strategies for crystalline solar cell recycling: A review on recycling techniques, companies, and environmental impact analysis. *Sustainability*, 16(13):5785, 2024.

AT Alabi and MO Jelili. Clarifying Likert scale misconceptions for improved application in urban studies. *Quality & Quantity*, 57(2):1337–1350, 2023.

E Aldarsi and R Singh. Review of community solar in the United States of America. *New Energy Exploitation and Application*, 3(1):67–88, 2024.

AO Amole, S Oladipo, OE Olabode, KA Makinde, and P Gbadega. Analysis of grid/solar photovoltaic power generation for improved village energy supply: A case of Ikose in Oyo State Nigeria. *Renewable Energy Focus*, 44:186–211, 2023.

Y Chen, D Chen, PP Altermatt, S Zhang, L Wang, X Zhang, J Xu, Z Feng, H Shen, and PJ Verlinden. Technology evolution of the photovoltaic industry: Learning from history and recent progress. *Progress in Photovoltaics: Research and Applications*, 31(12):1194–1204, 2023.

M Dada and P Popoola. Recent advances in solar photovoltaic materials and systems for energy storage applications: A review. *Beni-Suef University Journal of Basic and Applied Sciences*, 12(1):1–15, 2023.

KAN Dokyi and A Sharifi. Identification and analysis of barriers to the implementation of utility-scale solar photovoltaic technology in Ghana. *Energy for Sustainable Development*, 83:101547, 2024.

N Fauzi and FZ Ismail. Raising awareness and overcoming challenges: Recommendations for Solar Photovoltaic (PV) system adoption in residential areas. *International Journal of Business and Technology Management*, 6(S2):278–293,2024.

M Formolli, G Lobaccaro, and J Kanters. Solar energy in the Nordic built environment: Challenges, opportunities and barriers. *Energies*, 14(24):8410, 2021.

S Freitas, M Silva, E Silva, A Marceddu, M Miccoli, P Gnatyuk, L Marangoni, and A Amicone. An artificial intelligence-based framework to accelerate data-driven policies to promote solar photovoltaics in Lisbon. *Solar RRL*, 7(24):2300597, 2023.

MN Hussain, MR Zaman, MA Halim, MS Ali, and MYA Khan. A comprehensive review of renewable and sustainable energy sources with solar photovoltaic electricity advancement in Bangladesh. *Control Systems and Optimization Letters*, 2(1):1–7, 2024.

MC Ibegbulam, OO Adeyemi, and OC Fogbonjaiye. Adoption of solar PV in developing countries: Challenges and opportunity. *International Journal of Physical Sciences Research*, 7(1):36–57, 2023.

O Ideki, B Barikor, and O Ajoku. Assessment of natural resources for energy transition in Rivers State, Nigeria. *Discover Energy*, 4(1):11, 2024.

MW Ijeoma, H Chen, M Carbajales-Dale, and RO Yakubu. Techno-economic assessment of the viability of commercial solar PV system in Port Harcourt, Rivers State, Nigeria. *Energies*, 16(19):6803, 2023.

OO Ilesanmi, DT Moyanga, AE Oke, and J Aliu. Unraveling the awareness dynamics of smart building features and technologies in the Nigerian construction sector. *International Journal of Building Pathology and Adaptation*, 2024. https://doi.org/10.1108/IJBPA-04-2024-0073.

OO Ilesanmi, AE Oke, DT Moyanga, JO Aliu, and LD Ojo. Innovative strategies to prevent flooding in buildings and infrastructure in Nigeria. *International Journal of Disaster Resilience in the Built Environment*, 16(5):688–710, 2025. https://doi.org/10.1108/IJDRBE-03-2025-0040.

NC Kayaçetin and B Hozatlı. Whole life carbon assessment of representative building typologies for nearly zero energy building definitions. *Journal of Building Engineering*, 95:110214, 2024.

P Kumar, P Kumar, and V Aggarwal. A behavioral study on villagers' adoption intention and carbon neutrality toward rooftop solar photovoltaic systems in India. *International Journal of Energy Sector Management*, 18(1):200–227, 2024.

C Li and C Chen. Single-crystal perovskite for solar cell applications. *Small*, 20(48):2402759, 2024.

HNN Manuel, HM Kehinde, CP Agupugo, and ACN Manuel. The impact of AI on boosting renewable energy utilization and visual power plant efficiency in contemporary construction. *World Journal of Advanced Research and Reviews*, 23(2):1333–1348, 2024.

I Mashal, OA Khashan, M Hijjawi, and Alshinwan, M. The determinants of reliable smart grid from experts' perspective. *Energy Informatics*, 6(1):10, 2023.

RM May, KH Goebbert, JE Thielen, JR Leeman, MD Camron, Z Bruick, EC Bruning, RP Manser, SC Arms, and PT Marsh. MetPy: A meteorological Python library for data analysis and visualization. *Bulletin of the American Meteorological Society*, 103(10):2273–2284, 2022.

A Mishrif and A Khan. Non-industrial solar energy use, barriers, and readiness: Case study of Oman. *Energies*, 17(16):3917, 2024.

M Mogrovejo-Narvaez, A Barragán-Escandón, E Zalamea-León, and X Serrano-Guerrero. Barriers to the implementation of on-grid photovoltaic systems in Ecuador. *Sustainability*, 16(21):9466, 2024.

M Mustafa, MOF Malik, and A Maqsoom. Barriers to solar PV adoption in developing countries: Multiple regression and analytical hierarchy process approach. *Sustainability*, 16(3):1032, 2024.

U Mutumbi, G Thondhlana, and S Ruwanza. Adoption of residential rooftop solar PV systems in South Africa: A scoping review of barriers. *Heliyon*, 10(10):1–16, 2024.

JNL Naraharisetti, R Devarapalli, and V Bathina. Parameter extraction of solar photovoltaic module by using a novel hybrid marine predators–success history based adaptive differential evolution algorithm. *Energy Sources, Part A: Recovery, Utilization, and Environmental Effects*, 46(1):14550–14572, 2024.

M Numata, M Sugiyama, and G Mogi. Barrier analysis for the deployment of renewable-based mini-grids in Myanmar using the analytic hierarchy process (AHP). *Energies*, 13(6):1400, 2020.

R Nyffenegger, Ä Boukhatmi, T Radavičius, and M Tvaronavičienė. How circular is the European photovoltaic industry? Practical insights on current circular economy barriers, enablers, and goals. *Journal of Cleaner Production*, 448(24):141376, 2024.

CP Ohanu, GN Egbo, and T Sutikno. Sizing and analysis of a standalone photovoltaics system for a three-bedroom residence in Nigeria. *Indonesian Journal of Electrical Engineering and Computer Science (IJEECS)*, 33(1):1, 2024.

AE Oke, J Aliu, TM Akinpelu, OO Ilesanmi, and KT Alade. Breaking barriers: Unearthing the hindrances to embracing energy economics principles in Nigerian building projects. *Energy and Built Environment*, 6(3):534–544, 2025.

BC Okpala and CC Nzeanorue. Smart energy management in Nigeria: Implementing IoT and AI for sustainable urban development. *Path of Science*, 10(10):2012–2020, 2024.

SALU Olaleye, I Ogundari, ANIH Samuel, and JK Akinbami. Assessment of roof tilt and building azimuth for off-grid photovoltaic power for buildings in metropolitan Lagos, Nigeria. *Journal of Digital Food, Energy & Water Systems*, 5(1):1–6, 2024.

G Ongeri and P Mbataru. Determinants of adoption of solar energy for home use in Nairobi City County, Kenya. *International Academic Journal of Arts and Humanities*, 1(3):167–183, 2023.

IC Onwuadiochi, MA Ijioma, JO Mage, and ML Ozoemene. Comparative analysis of the measured and extrapolated wind shear distributions at Murtala Mohammed and Port Harcourt international airports, Nigeria. *Journal of Environment and Earth Science*, 11(5):63–78, 2021.

IC Osuizugbo, AA Orekan, MM Omer, and RF Simon. An exploratory factor analysis on technological-related barriers to the construction of zero-energy buildings in Nigeria. *International Journal of Building Pathology and Adaptation*, Vol. ahead-of-print No. ahead-of-print, 2024.https://doi.org/10.1108/IJBPA-05-2024-0091.

M Pal, R Dass, and V Nehra. Barrier identification and prioritisation for the deployment of solar photovoltaic systems in dairy farming in India using the ISM–MICMAC approach. *International Journal of Sustainable Energy*, 42(1):152–181, 2023.

C Parsad, S Mittal, and R Krishnankutty. A study on the factors affecting household solar adoption in Kerala, India. *International Journal of Productivity and Performance Management*, 69(8):1695–1720, 2020.

S Raschka, J Patterson, and C Nolet. Machine learning in Python: Main developments and technology trends in data science, machine learning, and artificial intelligence. *Information*, 11(4):193, 2020.

Z Ren, Y Chen, C Song, M Liu, A Xu, and Q Zhang. Economic analysis of rooftop photovoltaics system under different shadowing conditions for 20 cities in China. In *Building Simulation*, (Vol. 17, No. 2, pp.235–252). Beijing, Tsinghua University Press, 2024.

IK Rotich, H Chepkirui, and PK Musyimi. Renewable energy status and uptake in Kenya. *Energy Strategy Reviews*, 54:101453, 2024.

R Roy and JM Pearce. Is small or big solar better for the environment? Comparative life cycle assessment of solar photovoltaic rooftop vs. ground-mounted systems. *The International Journal of Life Cycle Assessment*, 29(3):516–536, 2024.

C Sheng and J Liu. A systematic analysis of stakeholder interaction and the barriers towards upscaling urban residential solar PV: The case of Shanghai, China. *Energy Strategy Reviews*, 50:101259, 2023.

Q Shen, X Wen, S Xia, S Zhou, and H Zhang. AI-based analysis and prediction of synergistic development trends in US photovoltaic and energy storage systems. *International Journal of Innovative Research in Computer Science & Technology*, 12(5):36–46, 2024.

B Silveira, MEM dos Santos, FJF Maia, AP Basso, JN Singh, and HK de Medeiros Costa. Incentives for photovoltaic energy generation: A comparative analysis of policies in Spain, Germany, and Brazil. *Energy Strategy Reviews*, 54:101415, 2024.

RJ Stern, A Fernandez, EA Jacobs, TB Neilands, R Weech-Maldonado, J Quan, A Carle, and HK Seligman. Advances in measuring culturally competent care: A confirmatory factor analysis of CAHPS-CC in a safety-net population. *Medical Care*, 50:S49–S55,2012. https://doi.org/10. 1097/mlr.0b013e31826410fb.

H Sun, RU Awan, MA Nawaz, M Mohsin, AK Rasheed, and N Iqbal. Assessing the socio-economic viability of solar commercialization and electrification in South Asian countries. *Environment, Development and Sustainability*, 23(1):9875–9897, 2021.

P Thakur and VH Wilson. Analysis of barriers affecting the adoption of community solar from consumer's perspective: A hybrid ISM-DEMATEL approach. *Energy & Environment*, 35(1):113–141, 2024.

AH Tunde, SO Yusuf, AI Taiwo, G Ocran, P Owusu, and AH Paul-Adeleye. AI-driven innovations in energy efficiency: Transforming smart buildings and urban areas through technology and digital transformation. *World Journal of Advanced Research and Review*, 24(01):141–152, 2024. https://doi.org/10.30574/wjarr.2024.24.1.2921.

FO Usman, EC Ani, W Ebirim, DJP Montero, KA Olu-lawal, and N Ninduwezuor-Ehiobu. Integrating renewable energy solutions in the manufacturing industry: Challenges and opportunities: A review. *Engineering Science & Technology Journal*, 5(3):674–703, 2024.

H Van Tran, AV Tran, NQA Ho, and DN Pham. Factors influencing the decision to use rooftop solar power systems in Vietnam. *Journal of Infrastructure, Policy and Development*, 8(6):4631, 2024.

F Wang and W Liu. The current status, challenges, and future of China's photovoltaic industry: A literature review and outlook. *Energies*, 17(22):5694, 2024.

AH Xiao. Lagos in life: Placing cities in lived experience. *The Journal of Modern African Studies*, 59(3):391–412, 2021.

6

Predicting Comfort: AI-driven HVAC for Intelligent Energy Management

*Tuğçe Aker**

Product Department, Faradai Limited, London, UK

*Corresponding author: Aker T.; tugceaker@gmail.com

6.1 Challenge

Buildings account for approximately one-third of global energy consumption, with 38% of this attributed to heating, ventilation, and air conditioning (HVAC) systems (IEA, a). Despite their significant energy consumption, these systems often leave occupants dissatisfied (Karmann et al., 2018). The quest for thermal comfort in the built environment has become one of the most pressing challenges in sustainable energy management. As global temperatures rise and extreme weather events become more frequent (NASA), the energy demands for space conditioning—both heating and cooling—have reached unprecedented levels.

Since 2000, the global energy demand for space cooling has grown at a remarkable rate, increasing by approximately 4% annually—twice the rate of the demand for water heating. This growth is primarily driven by the explosive adoption of residential air conditioning, with global installations tripling to over 1.5 billion units by 2022 (IEA, b). In addition to cooling systems, space and water heating account for almost half of all building energy consumption (IEA, c). The challenges extend beyond cooling to heating systems, which remain heavily reliant on fossil fuels and constrained by outdated electricity infrastructure.

Sixty percent of the energy demand for heating systems is still met by fossil fuels. Natural gas dominates the heating landscape in major economies, including the United States, the European Union, and China, contributing substantially to both direct and indirect CO_2 emissions (IEA, c).

Renewable Energy Transition with Artificial Intelligence: Challenge-driven Solutions, First Edition. Nina Dethlefs and Joyjit Chatterjee. Published 2026 by John Wiley & Sons, Inc.

Heating systems contribute to emissions by burning fossil fuels, while cooling systems, despite operating on electricity, can also destabilize the grid, especially during heat waves, due to sudden spikes in demand, increasing the risk of failures. Both systems pose challenges for sustainability and energy resilience.

Reducing energy demand and emissions requires a fundamental transformation in how we approach space conditioning. This transformation involves not only transitioning away from fossil fuels but also reimagining how thermal comfort is managed in buildings. The solution lies in expanding the electrification of heating systems, developing networks powered by renewable energy, and implementing intelligent control systems that optimize energy use while maintaining occupant comfort.

The predicted mean vote (PMV) model, developed by Fanger through thermoregulation studies (Fanger, 1970), is the current standard for thermal comfort assessment. Widely adopted in both national and international standards, it estimates the average thermal comfort response of large populations based on six key variables: indoor air temperature, relative humidity, mean radiant temperature, airspeed, clothing insulation, and metabolic rate.

However, the PMV model assumes passive interaction with the environment, which fails to capture real-time variations in individual thermal states (Wang et al., 2018; Daum et al., 2011; Salamone et al., 2018). Traditional HVAC systems, which rely on fixed set points guided by the PMV model, enforce a one-size-fits-all thermal comfort standard. Yet, research over the past five decades has increasingly challenged this approach, highlighting significant variations in individual and contextual comfort preferences.

Moreover, the PMV model lacks real-time user-specific feedback, limiting its ability to adapt to dynamic environmental and physiological changes. As a result, HVAC systems based on PMV prioritize maintaining temperature and humidity within predefined thresholds rather than responding to real-time occupant needs.

Given these limitations, occupant-centric control (OCC) models have gained increasing attention as a more adaptive and responsive approach to HVAC management. OCC prioritizes occupant needs and preferences, integrating them into the operation of various building systems. By leveraging advanced monitoring technologies, OCC analyzes real-time data on indoor environmental quality, occupant presence, and interactions to optimize both energy efficiency and occupant well-being (Nagy et al., 2023; Wegertseder-Martínez, 2023; Feng et al., 2015).

Figure 6.1 illustrates the core dimensions of OCC, emphasizing the interplay between human experience and building systems. It highlights the various factors influencing thermal comfort, usability, and energy efficiency, demonstrating how OCC places occupants at the center of decision-making for a more adaptive and intelligent indoor climate control strategy.

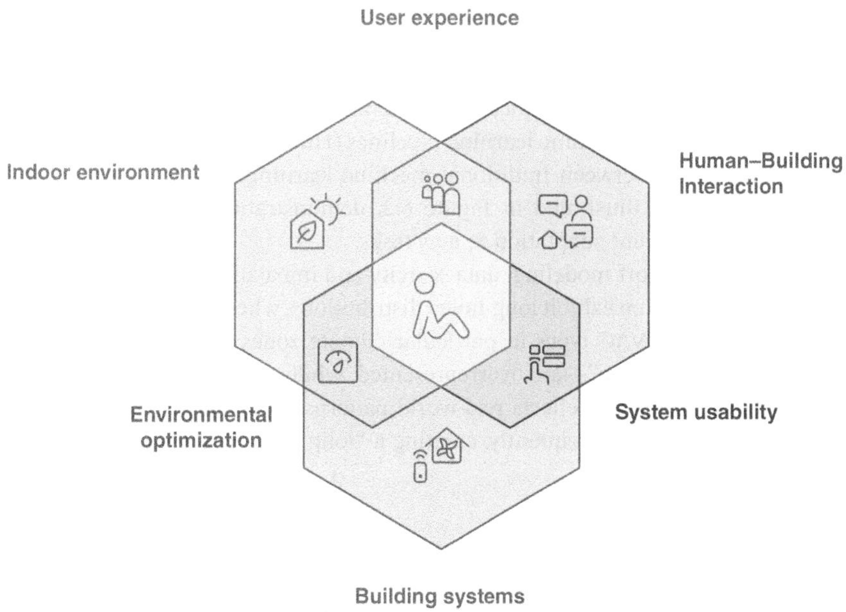

Figure 6.1 Occupant-centric control (OCC) framework.

These systems enable smoother integration with the energy grid, improve distribution efficiency, reduce emissions, and support the transition to renewable energy sources. Additionally, they help establish resilient energy infrastructures. By leveraging predictive algorithms and advanced control strategies, HVAC systems can dynamically adjust their operations in response to fluctuations in renewable energy availability and grid demands. This adaptability minimizes peak energy demand, balances loads, and optimizes energy storage utilization, ensuring HVAC operations are synchronized with energy supply variations.

Despite their significant potential to enhance sustainable building operations, data-driven approaches face limitations. Traditional machine learning models require a substantial amount of labeled data to learn and perform effectively. However, they operate in isolation on a single task, lacking the ability to transfer knowledge to new tasks. These limitations pose significant challenges for real-world applications, particularly in HVAC systems, where collecting large-scale labeled datasets is time-consuming and costly.

To address these challenges, transfer learning and meta-learning have been introduced. Transfer learning leverages labeled data or prior knowledge from related domains to enhance the performance of machine learning models in a specific domain. This technique is particularly powerful as it allows models to generalize across different but related tasks, thereby reducing the need for extensive labeled data (Pan, 2014; Zhuang et al., 2020).

Meta-learning, or "learning to learn," extends transfer learning by accumulating experience from multiple learning tasks and applying this knowledge to new tasks or domains. Unlike traditional models that start from scratch, meta-learning optimizes learning strategies, significantly accelerating model adaptation and improving the efficiency of machine learning pipelines (Hutter, 2019).

The differences between traditional machine learning, transfer learning, and meta-learning are illustrated in Figure 6.2, demonstrating how meta-learning enables more efficient adaptation to new tasks.

In thermal comfort modeling, data scarcity and imbalance present major challenges. Datasets often exhibit long-tailed distributions, where certain conditions—such as specific HVAC types in particular climate zones or thermal preference classes like "no change"—are overrepresented, while others remain underrepresented. This imbalance reflects real-world patterns, where some environmental conditions occur more frequently, creating a "long tail" of rare cases, as seen in Figure 6.3.

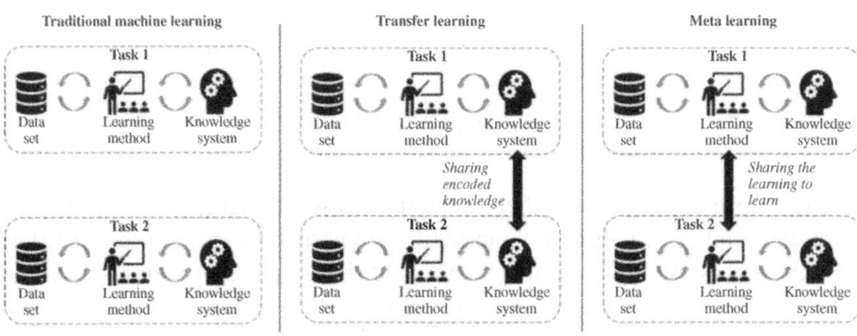

Figure 6.2 Comparison of traditional learning, transfer learning, and meta-learning (Koshiyama et al., 2020/with permission of Elsevier).

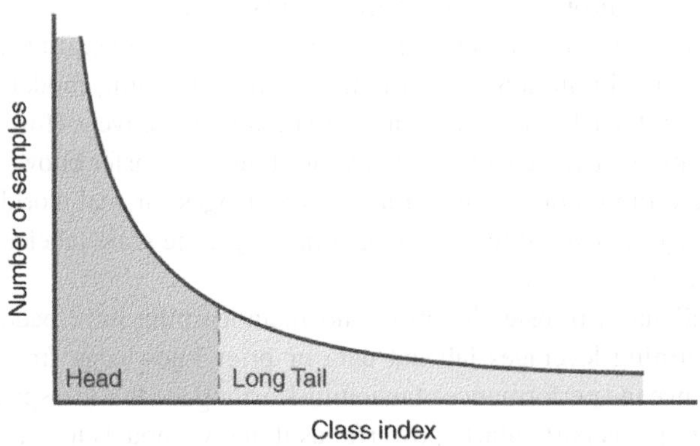

Figure 6.3 Real-world data distribution and meta-learning.

To mitigate these challenges, transfer learning and meta-learning have been applied to thermal comfort prediction (Chen et al., 2023; Zhang et al., 2024). These models can quickly adapt to new datasets by leveraging prior knowledge, making them particularly effective in scenarios with limited labeled data.

As shown in Figure 6.3, few-shot, one-shot, and zero-shot learning—collectively referred to as low-shot learning—allow models to extract meaningful patterns from minimal training examples. This is particularly beneficial in extreme data scarcity situations, while imbalanced classification methods help address the challenges posed by long-tailed distributions (Parnami and Lee, 2022).

The future of thermal comfort and HVAC energy management lies in the development of systems that dynamically adapt to both occupant-specific needs and sustainable energy goals. While challenges such as data scarcity, model generalization, and complex implementation persist, AI-driven HVAC solutions offer a promising path forward. By leveraging adaptive learning techniques and integrating real-time user feedback, these intelligent systems can redefine the balance between comfort and efficiency, paving the way for a more sustainable and occupant-centric built environment.

6.2 Case Study

This chapter introduces a novel approach to this challenge: AI-driven HVAC systems that bridge the divide between personal comfort and environmental sustainability. Using advanced predictive thermal comfort modeling, these systems can dynamically optimize building operations in real time, responding to both occupant preferences and grid conditions. This intelligent approach enables:

- Precise prediction and management of thermal comfort needs
- Dynamic load balancing to reduce peak demand stress
- Seamless integration with renewable energy sources
- Optimization of energy efficiency while maintaining comfort standards

By integrating artificial intelligence with traditional HVAC systems, we can transform buildings from passive energy consumers to active contributors to a sustainable energy future. In this section, we will outline the technical framework and implementation strategies for the proposed AI-driven system, illustrating how intelligent thermal management can help address the growing conflict between the demands for comfort and the need for environmental sustainability.

In this case study, we employ a meta-learning approach to classify thermal preferences, assessing the model's performance across unseen tasks and varying data availability. Through this analysis, we explore how AI can improve adaptability in HVAC operations, ensuring efficient and occupant-focused climate control even in situations with limited data.

6.2.1 Dataset Overview

A thermal comfort dataset is a collection of data related to the thermal conditions experienced by building occupants, including environmental parameters—such as indoor temperature and humidity—occupants' subjective responses, building system information, and occupants' personal details. These datasets are essential for understanding the thermal requirements of occupants to create energy-efficient building codes while maintaining comfort levels.

Open datasets especially play a crucial role in understanding and improving thermal comfort in various environments, particularly in building design and urban planning. The availability of extensive, high-quality datasets allows researchers to explore complex relationships between environmental variables and human comfort levels, leading to more effective design solutions and energy-efficient systems.

To focus specifically on thermal preference prediction, we selected five datasets from a larger pool based on the richness and quality of available thermal preference data. Each dataset captures a variety of environmental parameters, including indoor air temperature, relative humidity, climate and seasonal details, occupant characteristics (such as gender and BMI), and HVAC system types.

The dataset utilized in this research is a compilation of distinct open-source datasets designed to provide a representative sample across different climates, cultures, and personal influences on thermal preference. It consists of 97 762 data points from 32 countries, as illustrated in Figure 6.4. The dataset integrates environmental, occupant, and HVAC-related variables to predict thermal preference.

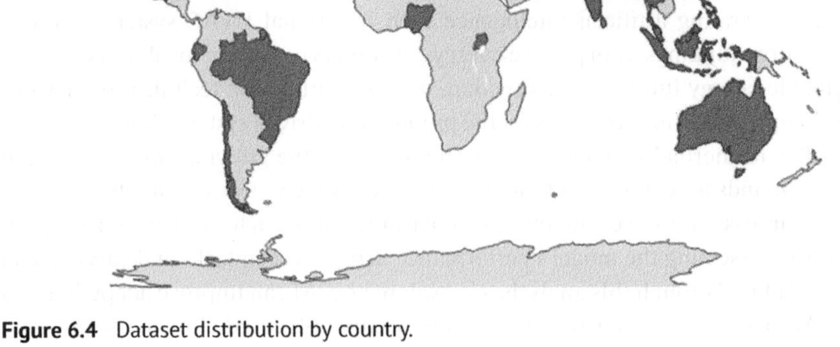

Figure 6.4 Dataset distribution by country.

The dataset contains the following primary features:

- **Indoor Environmental Parameters:** Indoor air temperature (°C), relative humidity (%), air velocity (m/s), mean radiant temperature (°C), and globe temperature (°C).
- **Outdoor Environmental Parameters:** Outdoor air temperature (°C), relative humidity (%), season, and climate class.
- **Building Information:** Location, building function, HVAC system type (naturally ventilated, air-conditioned, and mixed-mode).
- **Occupant Characteristics:** Metabolic rate (met), clothing insulation category (clo), gender, age, weight, and height.
- **Thermal Preference:** A three-class categorical variable indicating the occupant's desired temperature adjustment: "Cooler," "No Change," or "Warmer."

The specific datasets included in this study are as follows:

- ASHRAE Global Thermal Comfort Database (Thomas et al., 2022)
- Brazilian Indoor Thermal Comfort Database (Buonocore, 2022)
- Thermal Comfort Dataset – Darjeeling (Thapa, 2015)
- Field Study in Brisbane, Australia (Kramer et al., 2023)
- The Scales Project, a cross-national dataset on the interpretation of thermal perception scales (Schweiker et al., 2019)

The dataset represents three gross domestic product (GDP) categories and 20 distinct climate classes. The distribution details are visualized in Figure 6.5 and Table 6.1.

An analysis of the dataset reveals a significant class imbalance in thermal preference labels, where "No Change" responses dominate, while "Cooler" and "Warmer" preferences are underrepresented, as seen in Figure 6.6.

This imbalance could lead to the model being biased toward predicting the majority class, potentially reducing accuracy in scenarios where occupants desire a temperature change. Although techniques like oversampling and class-weighted loss functions are common solutions for addressing class imbalance, preliminary experimentation revealed that these methods provided minimal improvement in model performance while adding unnecessary complexity. To maintain model simplicity and computational efficiency without compromising predictive power, these traditional solutions were not pursued further.

By structuring tasks with seen and unseen conditions, the model's generalizability was tested to ensure it performs well, even when encountering underrepresented thermal preference labels. While the imbalance remains a challenge, the results indicate that the model can still capture thermal preference variations without requiring extensive dataset-balancing techniques.

Table 6.1 Number of samples per dataset.

Dataset	Records	Climate classes	Countries
ASHRAE Global Database	81 855	14	25
Brazilian Thermal Comfort	9095	3	1 (Brazil)
Scales Project	3380	18	20
Darjeeling Dataset	2619	2	1 (India)
Brisbane Field Study	813	1	1 (Australia)
Total	97 762	20	32

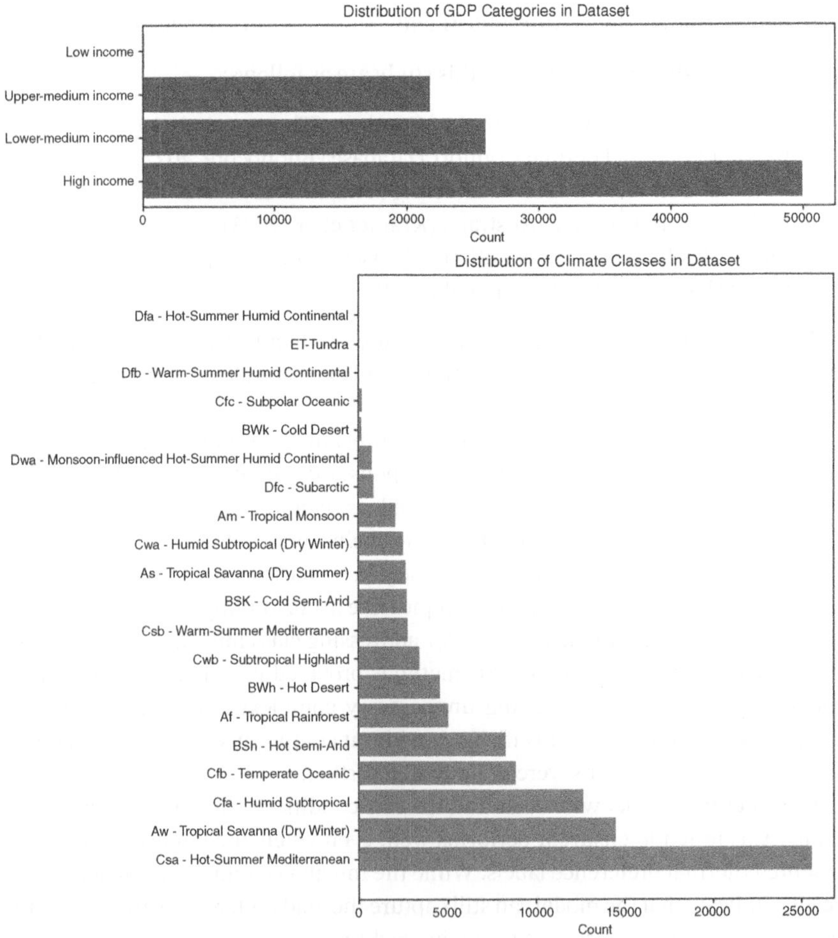

Figure 6.5 Dataset distribution based on GDP category and climate class.

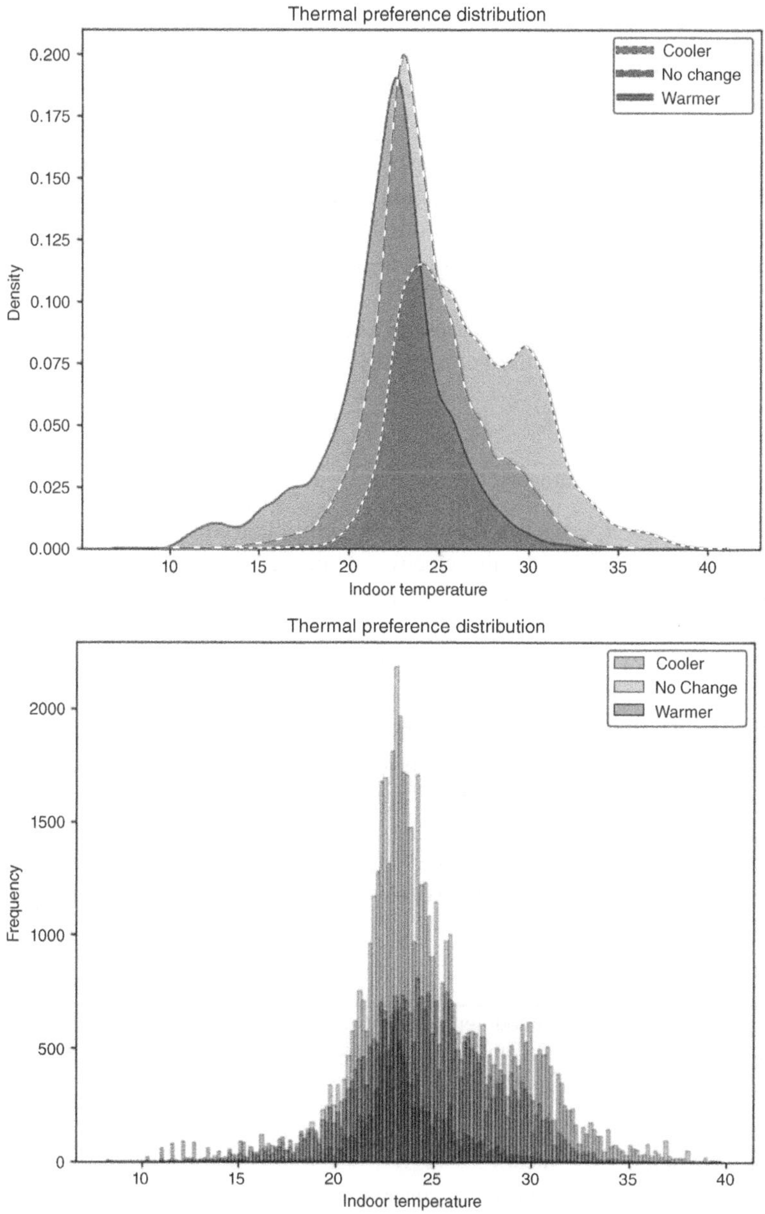

Figure 6.6 Thermal preference votes distribution.

6.2.2 Methodology

The meta-training strategy employed for thermal preference prediction aims to optimize for both accuracy and adaptability across environments with varying HVAC types, climate conditions, and demographic contexts. Conditional neural processes (CNPs) enable efficient learning by making precise predictions from just a few training data points, while also scaling effectively to complex functions and large datasets (Garnelo et al., 2018). Additionally, CNPs are computationally efficient and highly flexible, making them well-suited for real-time HVAC applications (Prokhorov et al., 2023).

To ensure model generalizability across different building types while maintaining practical applicability, we used only three input parameters:

- Indoor air temperature,
- Relative humidity,
- Clothing insulation category.

Unlike traditional approaches that require additional environmental inputs such as globe temperature and air velocity, our model effectively predicts thermal preference using standard HVAC-relevant parameters. This choice makes the model easier to implement in real-world scenarios since indoor temperature and humidity sensors are widely available in modern buildings.

For model architecture, we implemented a CNP with 2 layers and 64 neurons per layer. This structure balances efficiency and performance, enabling the model to learn meaningful representations of thermal preference while remaining computationally lightweight. The model architecture details are summarized in Table 6.2.

Our meta-learning approach involved creating diverse tasks by combining HVAC type, GDP category, and climate class. Each unique combination defined a

Table 6.2 Model architecture.

Input features	Indoor temperature, relative humidity, clothing insulation
Model type	Conditional neural process (CNP)
Number of layers	2
Neurons per layer	64
Activation function	ReLU
Output dimensions	3 (Thermal preference: cooler, no change, warmer)
Optimizer	Adam
Learning rate	0.001
Loss function	Cross-entropy loss
Training epochs	1000

specific thermal comfort scenario, allowing the model to efficiently learn to adapt to new, previously unseen situations. This adaptability is essential for HVAC systems, which must continuously respond to dynamic occupant behavior and varying environmental conditions.

Traditional comfort prediction algorithms typically predict average occupant responses, assuming static conditions and uniform populations. These methods lack real-time adaptability and fail to address individual and dynamic comfort needs effectively.

In contrast, the proposed CNP-based meta-learning approach was initially trained across five diverse datasets, encompassing a variety of scenarios. Generalization was then rigorously tested using previously unseen tasks, introducing new cultural and climatic dimensions. The results demonstrated strong adaptability and effectiveness in occupant-centric comfort management, highlighting the potential of meta-learning-driven HVAC control systems.

This meta-learning framework moves beyond traditional static methods, such as the PMV, offering an innovative, robust, and occupant-centric solution suitable for dynamic, energy-efficient building management.

6.2.3 Results

For each HVAC system type, GDP category, and climate class, 27 different tasks were created. Of these, 18 tasks were used for training and testing, while 9 tasks were reserved for evaluating model generalization capabilities on unseen data. The training and evaluation results are presented in Figure 6.7.

The meta-learning approach demonstrated strong capability in handling the long-tailed nature of thermal comfort data. Under frequent conditions (head of the distribution), the model achieved high accuracy of up to 78%, all while maintaining computational efficiency. In moderate-frequency scenarios (middle of the distribution), the model performed well with an accuracy around 63%.

With only three input parameters—indoor air temperature, relative humidity, and clothing insulation category—the model achieved an average accuracy of 63% on seen tasks and 58% on unseen tasks. These results highlight the model's strong generalization capabilities.

To address data scarcity, we also applied one-shot and few-shot learning to evaluate the model's performance on limited data. In this context, "shot" refers to the number of data points used for adaptation—e.g. 5-shot learning means training with only 5 data points, while 10-shot learning utilizes 10 data points. As shown in Figure 6.8, the model maintains similar performance levels even with reduced training data, demonstrating its adaptability to low-data scenarios.

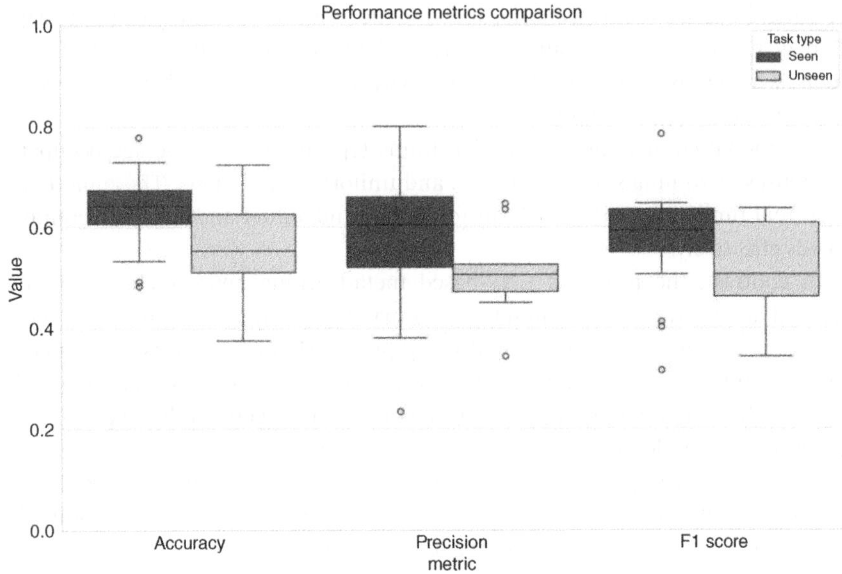

Figure 6.7 Model performance for seen and unseen tasks.

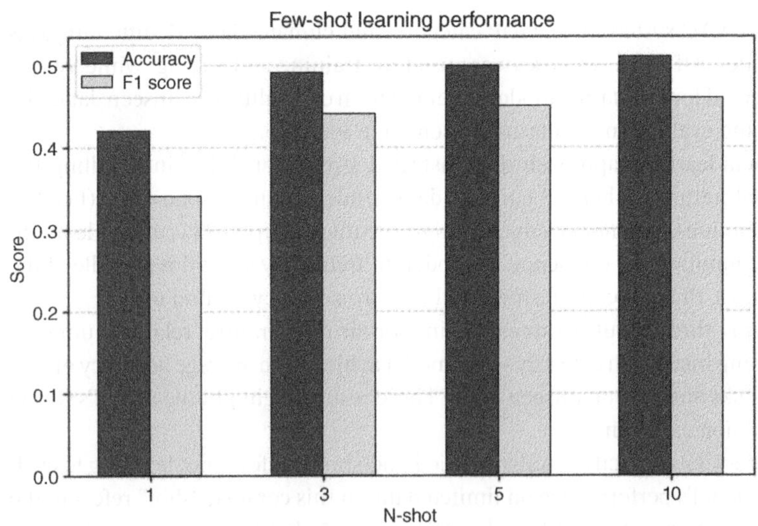

Figure 6.8 One-shot and few-shot learning performance for unseen tasks.

6.2.4 After

Traditional HVAC methods heavily relied on PMV-based models, lacking personalization in HVAC controls, often resulting in energy waste and occupant dissatisfaction. Data-driven approaches also have drawbacks, such as the requirement for substantial data. To overcome these limitations, the proposed meta-learning approach leverages diverse datasets to create a highly adaptable and occupant-focused HVAC control system. This system efficiently adapts to various conditions using fewer steps, enhancing scalability and practicality. As a result, this AI-driven framework rapidly aligns HVAC operations with occupant preferences, significantly improving comfort and reducing energy consumption across diverse building types and environmental conditions.

6.3 Discussion

The integration of AI-based models into HVAC systems represents a significant advancement in building management, offering the dual benefits of enhanced occupant comfort and reduced energy consumption. These models enable dynamic and precise adjustments tailored to the specific needs of occupants, contributing to the overall sustainability of building operations.

Unlike traditional PMV-based HVAC control, which relies on predefined thermal comfort assumptions, AI-driven systems continuously adapt to changing indoor conditions. This adaptability ensures that occupant comfort is maintained while minimizing energy waste, leading to smarter and more responsive building management.

While the meta-learning approach has demonstrated strong generalization capabilities, particularly in diverse HVAC environments, some challenges remain. The model performs well in common thermal preference categories but shows reduced accuracy in underrepresented conditions, such as extreme thermal discomfort scenarios. Additionally, integrating AI-driven HVAC into real-world infrastructure requires addressing computational efficiency, occupant acceptance, and potential regulatory constraints.

To further improve the proficiency and generalizability of these models, expanding data collection efforts is crucial. Incorporating a broader range of HVAC systems, climatic conditions, and geographic regions would enhance adaptability and minimize regional biases. This would ensure that AI-driven thermal comfort models remain robust and applicable to global applications.

Furthermore, integrating AI-driven HVAC control with smart grid infrastructure offers promising future directions. By dynamically adjusting HVAC operations based on renewable energy availability and real-time energy demand, AI models could contribute to load balancing and grid stability, reinforcing the link between energy efficiency and sustainability.

Despite these challenges, the adoption of AI-driven HVAC control systems marks a paradigm shift in energy-efficient building design. Future work should focus on improving self-learning capabilities, incorporating additional occupant feedback mechanisms, and refining real-time inference to enable seamless deployment in commercial and residential buildings.

6.4 Conclusion and Future Work

The integration of AI-driven prediction models into HVAC systems has paved the way for more personalized and efficient building climate control. By focusing on occupant-centric approaches, such as personal HVAC systems and individualized thermal comfort predictions, these advancements allow for improved determination of acceptable real-time HVAC setpoint intervals. This personalization not only enhances occupant comfort but also aligns HVAC operations with energy efficiency goals and real-time demand-response applications.

Future work should emphasize the following areas:

- Personalized HVAC Systems: Creating systems that address personal thermal preferences and sensitivities, allowing for targeted comfort while reducing energy consumption.
- Real-time Adaptability: Refining prediction models to dynamically adjust HVAC setpoints based on occupant behavior, environmental conditions, and energy grid demands.
- Data-driven Thermal Sensitivity Analysis: Expanding datasets to include diverse populations and settings to improve the accuracy and generalizability of thermal sensitivity predictions.
- Integration with Renewable Energy: Synchronizing HVAC operations with renewable energy availability, ensuring a seamless transition to sustainable energy infrastructures.

By addressing these areas, the potential for occupant-centric HVAC systems to revolutionize energy management and building sustainability can be fully realized.

Bibliography

Carolina Buonocore. Renata De Vecchi, Greici Ramos, Maira Andre, Christhina Candido, Maira Pires, Cesar Henrique de Godoy Gomes, and Roberto Lamberts. Brazilian indoor thermal comfort database. *Mendeley Data*, V1, 2022. https://data.mendeley.com/datasets/2ybzk7hbp2/1.

Liangliang Chen, Ayca Ermis, Fei Meng, and Ying Zhang. Meta-learning of personalized thermal comfort model and fast identification of the best personalized thermal environmental conditions. *Building and Environment*, 235:110201, 2023. https://doi.org/10.1016/j.buildenv.2023.110201.

David Daum, Frédéric Haldi, and Nicolas Morel. A personalized measure of thermal comfort for building controls. *Building and Environment,* 46(1):3–11, 2011. https://doi.org/10.1016/j.buildenv.2010.06.011.

Povl Ole Fanger. *Thermal Comfort: Analysis and Applications in Environmental Engineering*. p. 244. Copenhagen, Danish Technical Press, 1970.

Xiaohang Feng, Da Yan, and Tianzhen Hong. Simulation of occupancy in buildings. *Energy and Buildings*, 87:348–359, 2015. https://doi.org/10.1016/j.enbuild.2014.11.067.

Marta Garnelo, Dan Rosenbaum, Chris J Maddison, Tiago Ramalho, David Saxton, Murray Shanahan, Yee Whye Teh, Danilo J Rezende, and SM Ali Eslami. Conditional neural processes. In *International Conference on Machine Learning*, pp. 1704–1713. PMLR, 2018.

Frank Hutter, Lars Kotthoff, and Joaquin Vanschoren. *Automated Machine Learning: Methods, Systems, Challenges*. p. 219. Springer Nature, 2019. http://link.springer.com/10.1007/978-3-030-05318-5.

IEA. Buildings – topics, a. https://www.iea.org/topics/buildings.

IEA. Cooling – IEA buildings, b. https://www.iea.org/energy-system/buildings/space-cooling.

IEA. Heating – IEA buildings, c. https://www.iea.org/energy-system/buildings/heating.

Caroline Karmann, Stefano Schiavon, and Edward Arens. Percentage of commercial buildings showing at least 80% occupant satisfied with their thermal comfort. In *Proceedings of 10th Windsor Conference*, pp. 0–7. Windsor, UK, 2018. www.escholarship.org/uc/item/89m0z34x.

Adriano Koshiyama, Nick Firoozye, and Philip Treleaven. Algorithms in future capital markets. *Available at SSRN 3527511*, 2020.

T Kramer, V Garcia-Hansen, S Omrani, J Zhou, and D Chen. Personal differences in thermal comfort perception: Observations from a field study in Brisbane, Australia. *Building and Environment*, 245:110873, 2023. https://doi.org/10.1016/j.buildenv.2023.110873.

Zhongqi Miao and Ziwei Liu. Large-scale long-tailed recognition in an open World, 2019. http://bair.berkeley.edu/blog/2019/05/13/oltr/.

Zoltan Nagy, Burak Gunay, Clayton Miller, Jakob Hahn, Mohamed M Ouf, Seungjae Lee, Brodie W Hobson, Tareq Abuimara, Karol Bandurski, Maíra André, et al. Ten questions concerning occupant-centric control and operations. *Building and Environment*, 242:110518, 2023. https://doi.org/10.1016/j.buildenv.2023.110518.

NASA. Extreme weather - NASA science. https://science.nasa.gov/climate-change/extreme-weather/.

Sinno Jialin Pan. Transfer learning. In *Data Classification*, p. 34. Chapman and Hall/CRC, 2014.

Archit Parnami and Minwoo Lee. Learning from few examples: A summary of approaches to few-shot learning. *arXiv preprint arXiv:2203.04291*, 2022. http://arxiv.org/abs/2203.04291.

Victor Prokhorov, Ivan Titov, and N Siddharth. Autoencoding conditional neural processes for representation learning. *arXiv preprint arXiv:2305.18485*, 2023. http://arxiv.org/abs/2305.18485.

Francesco Salamone, Lorenzo Belussi, Cristian Currò, Ludovico Danza, Matteo Ghellere, Giulia Guazzi, Bruno Lenzi, Valentino Megale, and Italo Meroni. Application of IoT and machine learning techniques for the assessment of thermal comfort perception. *Energy Procedia*, 148:798–805, 2018. https://doi.org/10.1016/j.egypro.2018.08.130.

Marcel Schweiker, Amar Abdul-Zahra, Amar Abdul-Zahra, Maíra André, Farah Al-Atrash, Hanan Al-Khatri, Rea Risky Alprianti, Hayder Alsaad, Rucha Amin, Eleni Ampatzi, et al. The scales project, a cross-national dataset on the interpretation of thermal perception scales. *Scientific Data*, 6(1):289, 2019. https://doi.org/10.1038/s41597-019-0272-6.

Samar Thapa. Thermal comfort dataset - Darjeeling. *Mendeley Data*, V2, 2015. https://data.mendeley.com/datasets/55rddfmfsz/2.

Parkinson Thomas, Tartarini Federico, Földváry Ličina Veronika, Cheung Toby, Zhang Hui, de Dear Richard, Li Peixian, Arens Edward, Chun Chungyoon, Schiavon Stefano, et al. ASHRAE global database of thermal comfort field measurements. *Methods*, 2022:7–15, 2022. https://datadryad.org/stash/dataset/doi:10.6078/D1F671.

Zhe Wang, Richard De Dear, Maohui Luo, Borong Lin, Yingdong He, Ali Ghahramani, and Yingxin Zhu. Individual difference in thermal comfort: A literature review. *Building and Environment*, 138:181–193, 2018. https://doi.org/10.1016/j.buildenv.2018.04.040.

Paulina Wegertseder-Martínez. The need for a paradigm shift toward an occupant-centered environmental control model. *Sustainability*, 15(7):5980, 2023. https://doi.org/10.3390/su15075980.

Hejia Zhang, Seungjae Lee, and Athanasios Tzempelikos. Bayesian meta-learning for personalized thermal comfort modeling. *Building and Environment*, 249:111129, 2024. https://doi.org/10.1016/j.buildenv.2023.111129.

Fuzhen Zhuang, Zhiyuan Qi, Keyu Duan, Dongbo Xi, Yongchun Zhu, Hengshu Zhu, Hui Xiong, and Qing He. A comprehensive survey on transfer learning. *Proceedings of the IEEE*, 109(1):43–76, 2020. https://doi.org/10.1109/JPROC.2020.3004555.

7

Leveraging Generative AI-Driven Digital Twins for Renewable Energy Systems[1]

*Hanu Priya Indiran**

Department of Engineering, University of Cambridge, Cambridge, UK

*Corresponding author: Hanu Priya Indiran; hpi20@cam.ac.uk

7.1 Data and Communication Barriers

The global transition to renewable energy systems (RES) is essential for meeting sustainability goals and tackling climate change. Renewable energy sources such as solar, wind, hydro, and geothermal are at the forefront of worldwide decarbonization strategies. In contrast to fossil fuel-based systems, RES are dependent on natural energy sources that can be variable and unpredictable. This inherent variability introduces a level of complexity that necessitates innovative approaches to management, optimization, and scalability.

Artificial intelligence (AI) has emerged as a transformative force in this domain, offering solutions to critical challenges such as predicting energy output from fluctuating sources, coordinating distributed energy assets, maintaining grid stability, and facilitating predictive maintenance. Nonetheless, the integration of AI into renewable energy systems encounters significant obstacles. This chapter will explore two primary barriers that hinder its widespread adoption and effectiveness. Table 7.1 presents an organized overview of these challenges, categorizing them alongside their corresponding sub-challenges.

1 Chapter explores integrating generative AI in renewable energy systems.

Renewable Energy Transition with Artificial Intelligence: Challenge-driven Solutions, First Edition. Nina Dethlefs and Joyjit Chatterjee. Published 2026 by John Wiley & Sons, Inc.

Table 7.1 Key challenges in implementation of AI-driven renewable energy systems.

Challenge	Sub-challenges	Impact on AI implementation
Data scarcity	Limited historical data	Inadequate datasets limit AI's ability to accurately predict and optimize renewable energy system performance.
	Sparse datasets for rare but critical scenarios	Lack of representative data reduces AI model robustness in predicting critical, infrequent events.
	Heterogeneous and fragmented data sources	Complicates data integration, impairing AI model accuracy and operational effectiveness.
Critical communication infrastructure	Dynamic traffic loads and scaling issues	Limits the capability of AI-driven systems to perform real-time monitoring and control efficiently.
	Environmental stress affecting reliability	Communication disruptions compromise the reliability and responsiveness of AI-based solutions.
	Fault susceptibility over long distances	Impairs real-time data transmission, reducing the effectiveness of AI-driven predictive maintenance and optimization.

The Key Barriers:

1. **Data Scarcity:** AI models rely on comprehensive, high-quality datasets to train algorithms that facilitate predictive maintenance, optimize energy generation, and manage grid integration. These datasets must accurately reflect the complexity and variability of renewable energy systems, encompassing environmental fluctuations, equipment performance, and grid dynamics. Unfortunately, many renewable energy sources face considerable gaps in data availability and quality, which impede the development and implementation of effective AI models (Allal et al., 2024).
   - **Limited Historical Data:** Emerging renewable energy technologies, such as offshore wind farms and hybrid renewable systems, have limited operational histories. This lack of historical data restricts the ability to train AI models that can predict performance or optimize operations over time. For example, offshore wind farms in the North Sea face unique challenges like salt corrosion, high winds, and variable currents. Without sufficient historical datasets, it is difficult to model these dynamics accurately.

- **Sparse Datasets for Rare but Critical Scenarios:** Certain events, such as extreme weather conditions, grid failures, or equipment malfunctions, occur infrequently but have a disproportionate impact on system performance. AI models trained on typical operational data often fail to perform well under these rare conditions, which may include hurricanes affecting offshore wind farms or prolonged heatwaves reducing solar panel efficiency. Sparse data representing these scenarios limits the robustness of predictive models.
- **Heterogeneous and Fragmented Data Sources:** Renewable energy systems, particularly distributed energy resources (DERs), produce vast amounts of data from diverse sources, including IoT sensors, supervisory control and data acquisition (SCADA) systems, and external inputs like weather forecasts. These datasets are frequently fragmented, varying in formats, resolutions, and temporal scales. The challenge lies in harmonizing this dispersed data into a cohesive and reliable framework, as inconsistencies and inefficiencies in data handling can significantly degrade the performance of AI models designed to optimize and manage these systems.

2. **Challenges in Critical Communication Infrastructure:** Distributed renewable energy systems require robust, low-latency communication networks to enable real-time monitoring, control, and optimization. These networks facilitate the seamless flow of data between DERs (e.g. solar panels, wind turbines), grid operators, and AI-driven control systems, ensuring that decisions are made promptly and accurately. However, despite their critical role, these communication networks face several challenges that hinder the effective integration of AI into renewable energy systems (Liu, 2012).
   - **Dynamic Traffic Loads and Scaling Issues:** As distributed energy resources (DERs) expand, communication networks face the challenge of accommodating increasing data traffic while maintaining low latency (Liu, 2012). Optical networks, which are widely used for their high bandwidth, must manage:
     - *Traffic Spikes:* During peak energy production (e.g. midday for solar farms or high winds for wind turbines), data transmission demands surge. This creates congestion, especially when thousands of interconnected DERs are generating real-time performance metrics.
     - *Scalability Challenges:* As renewable energy systems grow, optical networks must scale to interconnect more DERs across wider areas. Ensuring consistent performance while scaling infrastructure remains a significant technical hurdle.

     For instance, a utility-scale solar farm with thousands of panels generates vast amounts of data, including performance metrics, temperature readings, and energy production statistics. Transmitting this data to a central AI-driven control system for analysis requires not only significant bandwidth but also efficient traffic management to prevent delays.

- **Sparse Datasets for Rare but Critical Scenarios:** Optical communication networks deployed in renewable energy systems often operate in harsh environments, where adverse conditions can significantly degrade their reliability (Uhl, 2022). In offshore wind farms, underwater optical cables are exposed to risks such as saltwater corrosion, strong ocean currents, and mechanical stress from seabed movement. Desert solar farms face challenges such as extreme heat, sandstorms, and UV radiation, which can deteriorate optical fibers and connectors over time. Additionally, extreme weather events, including storms, floods, and high winds, can damage both above-ground and underwater optical infrastructure, leading to service disruptions. Although optical networks are generally more robust than wireless systems, these environmental stressors increase the likelihood of physical damage and signal degradation, particularly in remote or hard-to-access locations.
- **Fault Susceptibility Over Long Distances:** Distributed renewable energy systems, such as offshore wind farms and long-distance solar grid integrations, rely heavily on optical networks to transmit data across vast distances. However, these networks face challenges that can compromise their reliability and efficiency. Signal quality diminishes over long distances due to fiber attenuation, requiring repeaters or amplifiers to maintain data integrity. Optical cables are also susceptible to mechanical stress and wear, caused by factors such as seabed movement, physical tension, or nearby construction activities (Uhl, 2022). Additionally, faults in underwater or remote optical cables are particularly challenging to diagnose and repair, often resulting in prolonged service disruptions. For instance, offshore wind farms depend on underwater fiber-optic cables to relay real-time data from turbines to onshore control centers. Any fault in these cables, whether from mechanical stress or environmental wear, can disrupt monitoring and optimization processes, delaying fault detection and increasing system downtime.

7.2 Case Study: The NorthWind Project

Offshore wind farms are at the forefront of the global shift toward renewable energy, utilizing robust and consistent wind resources over open waters to produce substantial electricity. This strategic deployment not only diversifies energy systems but also significantly reduces greenhouse gas emissions, making it a key pillar in achieving international climate goals. By the end of 2023, the global offshore wind capacity reached approximately 75 GW, with ambitious projections estimating a potential surge to 487 GW by 2033 (Global Wind Energy Council, 2024). This remarkable growth underscores the importance of optimizing offshore wind systems to meet the demands of a sustainable energy future.

AI plays a transformative role in offshore wind operations, enabling advanced capabilities such as predictive maintenance, energy forecasting, and autonomous operations. However, the integration of AI into offshore wind farms faces two critical barriers that this chapter aims to address: data scarcity and challenges in critical communication infrastructure. These obstacles not only impede the scalability of AI solutions but also limit their reliability in managing the complex and dynamic environments of offshore wind farms.

In this chapter, we present the NorthWind Project, a pioneering initiative in the North Sea, as a case study to explore how these barriers have traditionally hindered the adoption of AI in offshore wind farms (Rasheed et al., 2024). Supported by the Research Council of Norway, the project aims to drive down the costs of wind power while advancing sustainable development, placing a significant focus on the innovation and implementation of digital twin technologies. We analyze the limitations of earlier methods used to address data scarcity, highlighting their inefficiencies and gaps. The chapter then examines how the integration of Generative AI-driven Digital Twin technology has transformed this sector, addressing data scarcity and improving operational efficiency in the NorthWind Project.

Additionally, we examine the critical communication infrastructure challenges that have been flagged by the Connectivity Fieldlab North Sea, which has highlighted the significant reliability and scalability issues faced by offshore wind farms (NedZero, n.d.). Building on this, the chapter proposes extending the NorthWind Project by applying digital twin technology to critical communication infrastructure, particularly optical networks, to enhance their performance and reliability. This approach illustrates how such innovations can further support a seamless and robust AI transition for offshore wind farms, paving the way for a more resilient and efficient renewable energy future.

7.2.1 Conventional Practices: Tackling Data Scarcity

The challenges of data scarcity in offshore wind farms were traditionally managed using state-of-the-art methods that combined in-situ measurements, machine learning models, and data integration techniques. These approaches provided valuable insights into site conditions and operational planning but faced limitations due to high costs, sparse data availability, and reliance on interpolation, which introduced uncertainties.

1. **In-Situ Measurements and Interpolation:** Cone Penetrometer Tests (CPTs) were commonly employed to gather geotechnical data for offshore sites. However, these tests were costly, time-consuming, and provided data at only discrete locations, requiring interpolation to estimate seabed properties across a site. This approach often led to inaccuracies and uncertainties in data quality (Shoukat et al., 2023).
2. **Machine Learning Models:** Surrogate wave models, trained on limited measurement locations, were developed to predict wave conditions across broader spatial datasets, facilitating decision-making for installation and maintenance. While effective, these models relied heavily on existing data and lacked scalability for dynamic conditions (Whistler et al., 2024). Artificial neural networks (ANNs) were also used to extrapolate and optimize limited geotechnical data, helping to characterize offshore sites more effectively (Shoukat et al., 2023).
3. **Data Integration and Analytics:** Cloud-based platforms were utilized to integrate diverse datasets, such as wind resource outputs and geological information. This integration supported early insights and planning but was often decoupled from site-specific, ground-related conditions, limiting its robustness for dynamic offshore environments (Smith et al., 2023). Historical records and geo-data further aided in derisking development projects by providing foundational insights, though these datasets were often fragmented and lacked granularity. While these methods addressed some aspects of data scarcity, they struggled with key challenges:

   High Costs and Limited Scalability: In-situ measurements and machine learning models required extensive resources and were not easily scalable for large or complex offshore environments. Inconsistent

   Data Quality: Reliance on interpolation and sparse datasets introduced uncertainties, reducing the reliability of AI-driven optimization.

   Fragmentation of Insights: The decoupling of diverse datasets limited their ability to provide integrated, dynamic insights for real-time decision-making.

7.2.2 Conventional Practices: Reliability of Critical Communication Infrastructure

The reliability of critical communication infrastructure in offshore wind farms and distributed renewable energy systems has historically been managed through traditional engineering practices, redundancy strategies, and advanced network architectures. These approaches were designed to ensure continuous operation and minimize downtime, which are essential for the economic viability and operational efficiency of renewable energy systems. However, while effective to an extent, these methods were often resource-intensive, complex, and limited in scalability.

1. **Redundant Network Topologies:** To enhance reliability, offshore wind farms commonly employed redundant network topologies. High-performance rings (HiPERRings), redundant switches, and routers formed the backbone of communication networks, ensuring uninterrupted operation even in the event of link failures (Uhl, 2022). Alternative transmission technologies, such as WLANs (IEEE 802.11h, IEEE 802.11g), served as backup communication channels, further bolstering system resilience.
2. **Advanced Communication Architectures:** Ethernet passive optical networks (EPON) were proposed as a cost-effective solution for offshore communication networks, reducing complexity by eliminating active elements between optical network units (ONUs) and the central optical line terminal (OLT) (Ahmed et al., 2016). Protection schemes, such as bus-based and star-based EPON architectures, were evaluated to ensure data delivery even in the case of device or link failures (Ahmed and Kim, 2013). While these architectures enhanced reliability, they required careful planning and significant investment to meet offshore deployment standards.
3. **Basic Fault Detection Systems and Scheduled Maintenance:** Tools like optical time-domain reflectometers (OTDR) were used to locate faults along optical cables (El-Sayed et al., 2010). OTDRs provided only static snapshots, requiring system downtime for testing and limiting their usefulness in dynamic environments where real-time insights are crucial. Periodic inspections and maintenance routines were conducted to identify and repair damaged optical fibers, connectors, or amplifiers. This approach relied heavily on human intervention, particularly for underwater cables, and often resulted in reactive maintenance, addressing faults only after they caused significant disruptions (Cruzes, 2023). Distributed optical fiber sensors were deployed to monitor dynamic cables, providing real-time insights into mechanical, electrical, and thermal conditions and enabling preemptive failure mitigation (Song et al., 2023).
4. **Environmental Stress Modeling:** Physics-based predictive models estimated the impact of environmental factors such as ocean currents, saltwater corrosion, and temperature fluctuations on optical cables (Sahoo et al., 2019). These models were computationally expensive, lacked scalability, and were unsuitable for dynamic, real-time operations.

While these methods provided a robust framework for ensuring communication reliability, they faced several limitations: *High Costs:* Redundant infrastructure and industrial-grade components significantly increased deployment and maintenance costs. *Complexity:* Designing and maintaining advanced network architectures required specialized expertise and extensive planning. *Reactive Maintenance:* Traditional systems often depended on periodic inspections or fault-triggered responses, leading to inefficiencies and potential downtime. *Limited Scalability:* Expanding these systems to accommodate the growing size and complexity of offshore wind farms posed technical and economic challenges.

7.2.3 Methodology: Generative AI-driven Digital Twin Framework

The NorthWind Project showcases the transformative potential of generative AI-driven digital twins in offshore wind farms. By integrating generative AI with digital twin technology, it effectively addresses the challenge of data scarcity. The framework combines real-time data acquisition, synthetic data generation, and efficient computational modeling to deliver robust, scalable, and adaptive solutions for wind farm operations.

Digital Twin

A digital twin is a dynamic virtual representation of physical systems or processes that enables bidirectional information flow between the physical and virtual models. It utilizes data and simulation tools to facilitate real-time monitoring, forecasting, optimization, and decision-making (Attaran and Celik, 2023). Implementing digital twins involves three key steps: data acquisition, data modeling, and data application (Lv and Xie, 2022). Figure 7.1 illustrates a framework structure for a digital twin designed for comprehensive life-cycle management. This framework integrates products and manufacturing systems into a unified architecture, enhancing a three-dimensional digital twin model with additional data and services. It supports alternating execution of physical and virtual processes throughout the life cycle and extends into a five-dimensional digital twin model, encompassing physical objects, the twin model, services, twin data, and their interactions, as depicted in the outer operational layer (Tao et al., 2024).

The framework in Figure 7.1 emphasizes the pivotal role of data and reliable communication infrastructure in the effective implementation of digital twins, particularly for life-cycle management. Data forms the backbone of the digital twin framework, enabling precise modeling, analysis, and decision-making across both physical and virtual domains throughout the life cycle of renewable energy systems. However, ensuring the seamless flow of high-quality data across the various phases—design, deployment, operation, and decommissioning—remains a significant challenge.

Equally critical is the reliability of communication infrastructure, which facilitates interaction and bidirectional communication between physical systems and

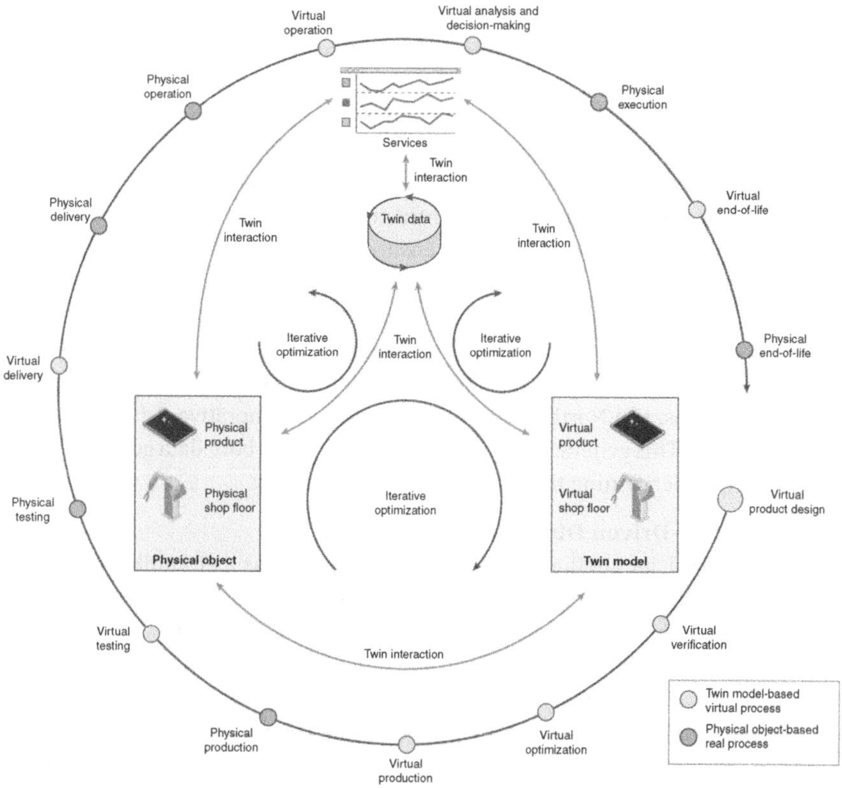

Figure 7.1 Conceptual model of a digital twin illustrating bidirectional data exchange between physical assets and virtual representations, supporting real-time monitoring, iterative optimization, and life-cycle management (Tao et al., 2024 / Springer Nature).

their digital counterparts at every stage of the life cycle. Disruptions in communication can undermine real-time monitoring, optimization, and control, thereby reducing the digital twin's efficacy in life-cycle management. In the NorthWind Project, a structured hierarchy with levels ranging from 0 to 5 was developed to assess digital twin capabilities. These levels advance from basic standalone functionality (Level 0) to descriptive and diagnostic features (Levels 1 and 2), predictive and prescriptive insights (Levels 3 and 4), and ultimately autonomous operation at Level 5 (Rasheed et al., 2024).

The digital twin was implemented through three key processes: data acquisition and generation, data standardization and management, and computationally efficient modeling. Data acquisition and generation utilized reverse engineering techniques and synthetic data creation from structural simulations and high-fidelity modeling of wakes and complex terrain scenarios. Data standardization and management leveraged a high-detail asset information model to manage and correlate diverse data streams. The asset layer encompassed asset-specific data—such as geometry, surface characteristics, and component temperature—alongside environmental data like weather conditions, terrain topology, and wind characteristics. This framework also defined data access and transfer layers, which are vital for managing critical communication infrastructure.

Computationally efficient modeling ensured the real-time functionality of the digital twin by employing reduced-order modeling techniques. These methods projected computationally intensive full-order models into reduced-dimensional spaces using orthogonal decomposition, enabling real-time simulation (Lucia et al., 2004). To generate synthetic, real-time atmospheric turbulent flow data for complex terrains, the NorthWind Project employed Generative Adversarial Networks (GANs). This generative AI-based approach drove both data generation and simulation, underpinning the digital twin's functionality.

Generative AI-Driven Digital Twins

Generative artificial intelligence (GenAI) has emerged as a groundbreaking field and has revolutionized content generation and synthetic data creation, showcasing its unique ability to create novel content from realistic data. By leveraging sophisticated machine learning algorithms and neural networks, generative AI generates new data samples similar to a given set of input data, driving innovation in numerous fields. Key models driving this innovation include generative adversarial networks (GANs), which use adversarial training between a generator and discriminator to produce realistic outputs and variational autoencoders (VAEs), which encode input data into a latent space to enable smooth data interpolation and diverse synthetic data generation (Creswell et al., 2018). Transformers, such as GPT, employ self-attention mechanisms to generate coherent text and other sequential data, while Diffusion Models iteratively refine noise into structured outputs, excelling at detailed image synthesis and complex data generation (Cao et al., 2022; Zhang et al., 2023). Recent advancements have further expanded the scope of generative AI, particularly with the development of large language models (LLMs) like GPT-4, which enhance synthetic text generation for applications such as virtual assistants and automated content creation. Multimodal models that process and generate text, images, and audio simultaneously open up new possibilities for complex datasets (Suzuki and Matsuo, 2022). Using generative AI models to leverage data generation to combat data scarcity for the efficiency of digital twins has been a recent transformative development, which is explored across various sectors, especially in urban digital twins focused on smart cities (Xu et al., 2024).

The NorthWind Project uses a class of Generative AI model, the GANs to generate synthetic real-time data, hence driving the digital twin's simulations. In the development of digital twin for offshore wind farms, one crucial data required is the turbulent flow data based on atmospheric conditions in various terrain conditions. As these flows are characterized by a wide spectrum of spatial and temporal variations, it is computationally highly demanding to simulate real-time data for complex terrains. To overcome this limitation, GANs are used to generate synthetic data while maintaining the accuracy typically associated with high-resolution models, where GANs are used to upscale coarse-resolution simulation data to finer resolutions (Rasheed et al., 2024). Figure 7.2 illustrates the working of the GANs framework to drive the digital twin (Rasheed et al., 2024).

GAN Architecture and Functionality

GANs are composed of two neural networks—the generator and the discriminator—that interact in a competitive learning process, leading to the production of highly realistic synthetic data:

1. *Generator:* The generator's primary objective is to synthesize realistic data from random noise or lower-resolution inputs. It typically employs deep convolutional neural networks (CNNs), composed of multiple layers, including convolutional, transposed convolutional (also known as deconvolutional), and activation layers (e.g. ReLU or Leaky ReLU). The generator receives input data—often noise vectors or coarse-resolution data—and progressively transforms these inputs through a series of layers into higher-resolution, realistic synthetic outputs. During training, it continuously refines the generated data based on the feedback received from the discriminator.

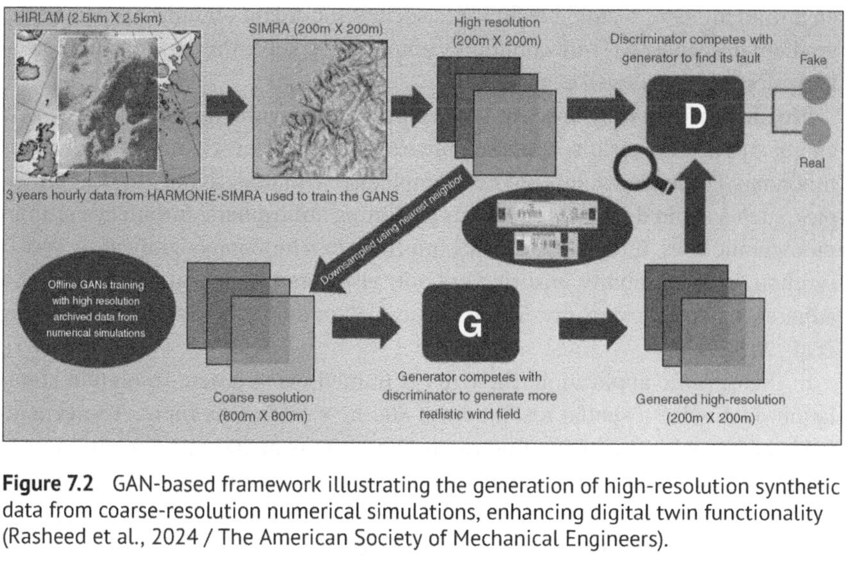

Figure 7.2 GAN-based framework illustrating the generation of high-resolution synthetic data from coarse-resolution numerical simulations, enhancing digital twin functionality (Rasheed et al., 2024 / The American Society of Mechanical Engineers).

2. *Discriminator:* The discriminator serves as a binary classifier, designed also using deep CNN architectures, comprising convolutional layers, activation functions, batch normalization, and fully connected layers. Its primary task is to differentiate real datasets from the generator's synthetic outputs. By assigning probabilities to input data (high probability for real data and low probability for synthetic data), the discriminator guides the generator's improvement by penalizing unrealistic outputs.

The GAN training is formally defined by a minimax optimization process, expressed mathematically as:

$$\min_{G} \max_{D} V(D, G) = \mathbb{E}x \sim Pdata(x)[\log D(x)] + \mathbb{E}_{z \sim p_Z(z)}[\log(1 - D(G(z)))] \quad (7.1)$$

Creswell et al. (2018), where:

- *G*: Generator network.
- *D*: Discriminator network.
- *Pdata*(x): Probability distribution representing real-world observed data.
- *Pz*(Z): Probability distribution from which the random noise vectors z are sampled, typically Gaussian.
- x: Real-world data samples.
- $G(z)$: Synthetic data generated from the noise vector z.

During training, the discriminator maximizes this objective by accurately distinguishing real data from generated samples, while the generator minimizes the same objective to fool the discriminator, creating an iterative adversarial learning environment. Optimization techniques such as the Adam optimizer or stochastic gradient descent (SGD) are commonly employed to solve this optimization problem effectively (Creswell et al., 2018).

Physics-informed GANs: An important technical enhancement in renewable energy applications is incorporating physics-informed constraints into GAN loss functions. These constraints enforce compliance with underlying physical principles, such as fluid dynamics equations governing atmospheric turbulence or thermodynamic laws. Integrating physics-informed loss functions significantly boosts synthetic data credibility, ensuring not only visual and statistical realism but also adherence to physical reality, which is essential for digital twin simulations (Yang et al., 2020).

In this specific application, the GAN is trained using coarse-resolution simulation data, with a spatial resolution of 800 m × 800 m, as input. The generator is tasked with producing fine-resolution outputs at 200 m × 200 m, thereby bridging the gap between computational efficiency and precision. A unique feature of this approach is the inclusion of a physics-informed cost function, which ensures that the outputs not only visually resemble high-resolution data but also adhere to the fundamental principles governing atmospheric flows (Rasheed et al., 2024). This aspect sets the method apart from conventional image upscaling techniques, which often disregard domain-specific constraints. The workflow for data upscaling and integration involves several key steps, starting with the preparation of low-resolution simulation data at 800 m × 800 m, obtained through numerical modeling techniques, which is then processed and fed into the GAN framework as input. During the training phase, the generator creates high-resolution fields while the discriminator evaluates their authenticity by comparing them to actual high-resolution data. A nearest neighbor algorithm is applied to the synthetic 200 m × 200 m high-resolution data to generate corresponding coarse-resolution inputs, ensuring consistency throughout training. Once trained, the GAN transforms incoming coarse-resolution simulation data into high-resolution outputs in real-time, enabling efficient and accurate modeling of atmospheric flows. The high-resolution data generated by the GAN is integrated with the digital twin, providing it with detailed spatio-temporal information for precise modeling and predictive analytics. By supplying data that closely mirrors real-world conditions, the GAN effectively addresses data scarcity, allowing the digital twin to perform reliably even in scenarios with limited direct observations or measurements.

Why GANs?

In the context of Generative AI-driven Digital Twins, GANs hold a unique position among various synthetic data generation approaches due to their technical advantages, especially when dealing with renewable energy systems.

GANs specifically excel because they use adversarial training, involving two neural networks—a generator and a discriminator—that competitively improve each other's performance. This ensures the synthetic data closely mirrors the statistical complexity of real-world observations, essential for accurate and reliable digital twins. In particular, GANs are capable of modeling rare but critical operational scenarios, such as extreme weather conditions and infrequent equipment malfunctions, thereby enhancing predictive robustness where traditional machine learning techniques might fail.

Compared to alternatives such as VAEs and diffusion models, GANs offer significant benefits. VAEs, while efficient for creating diverse synthetic data, typically generate lower fidelity outputs due to their reliance on probabilistic encoding into a latent space. Diffusion Models provide high-quality results, especially in image generation, but their computational complexity due to iterative refinement processes renders them less suitable for real-time applications, particularly in high-resolution, large-scale simulations required for dynamic renewable energy environments.

Moreover, GANs integrate physics-informed loss functions effectively, a feature critical for renewable energy applications. These physics-informed GANs ensure that generated data not only appears realistic but also adheres to the underlying physical and environmental constraints, essential for accurate digital twin simulations (Yang et al., 2020). This approach provides a considerable advantage over purely data-driven methods, which may produce unrealistic or physically implausible scenarios.

Additionally, GANs efficiently scale through resolution upscaling, allowing for a rapid transition from coarse-resolution simulations to fine-grained, high-resolution outputs. As demonstrated in the NorthWind Project, GANs can swiftly transform coarse-resolution data (800 m × 800 m) into fine-grained, high-resolution (200 m × 200 m) datasets, dramatically reducing computational load while maintaining accuracy. This scalability is crucial for large-scale renewable energy system modeling, enabling real-time simulation and proactive management.

Thus, GANs provide the optimal balance of realism, scalability, and computational efficiency, significantly outperforming alternative generative methods and making them particularly suited to addressing the data scarcity challenges faced in renewable energy system digital twin applications.

Digital Twin for Critical Communication Infrastructure of DES

In the context of Renewable Energy transitions with AI, particularly through the implementation of digital twins, achieving autonomy with bidirectional information flow is essential to fully leverage AI's potential in advancing distributed renewable energy systems. Critical communication infrastructure plays a vital role in enabling this data transfer, with offshore wind farms predominantly relying on optical fibers for networking (Dicorato et al., 2011; Liu, 2012; Uhl, 2022).

The Connectivity Fieldlab North Sea underscores the importance of assessing the reliability of these communication infrastructure assets to support digitally connected solutions. While the NorthWind Project models the data acquisition and data transfer layers within its digital twin framework, extending this model to incorporate critical communication infrastructure, such as optical networks, would significantly enhance its effectiveness. Including optical network systems as part of the asset layer would not only improve the overall digital twin framework but also enable detailed analysis of the reliability of these systems under the same atmospheric and environmental conditions simulated for wind farm operations.

By leveraging the generative AI techniques, such as GANs, already used to simulate environmental conditions and generate high-resolution data, these methods can be extended to model and evaluate the reliability of optical networks. This integration would provide valuable insights into the impact of communication infrastructure reliability on the efficiency and effectiveness of the digital twin, paving the way for more robust and autonomous renewable energy systems.

Integrating a digital twin for optical networks (DTON) framework presents an opportunity to enhance the reliability and efficiency of critical communication infrastructure in offshore wind farms. DTON encompasses three core components: real-time monitoring, mirror modeling, and automatic control (Wang et al., 2024). Measurement-informed models enhance the capability of DTONs by leveraging the capability of critical assets in the communication pipeline to generate data to build measurement-informed physical layer models in data transmission and reception (Faruk and Savory, 2024). Each of these components addresses unique challenges faced by optical networks in high-demand and environmentally harsh settings such as offshore wind farms. DTON framework, combined with Generative Adversarial Networks (GANs), can strengthen the communication infrastructure necessary for digital twins in offshore wind farms and distributed renewable energy systems.

1. **Real-time Monitoring and Data Integration:** Optical networks in offshore wind farm environments face unique challenges, including dynamic environmental variations, operational stresses, and high data throughput demands. Real-time monitoring is crucial for maintaining network reliability and ensuring uninterrupted data flow. Real-time data from optical fibers, amplifiers, and transceivers provide critical insights into the health and performance of the network (Faruk and Savory, 2024). Advanced tools like digital longitudinal monitoring (DLM) and pilot-tone-based monitoring allow for precise detection of environmental impacts and fault conditions in optical components (Faruk and Savory, 2024). These data feed directly into the digital twin, enabling continuous updates and proactive system adjustments. By reducing downtime, this integration ensures efficient operation of renewable energy systems.
2. **Mirror Modeling for Predictive Analytics:** Mirror modeling combines physics-based and data-driven approaches to simulate the behavior of optical networks under varying conditions. This predictive layer is vital for identifying potential failures and performance bottlenecks before they occur. Physics-informed machine learning (PIML) integrates the governing principles of optical signal propagation with real-time data inputs to create highly accurate predictions (Wang et al., 2024). Key optical network parameters, such as fiber attenuation profiles, signal-to-noise ratios, and nonlinear distortions, are modeled to ensure realistic simulation outcomes (Faruk and Savory, 2024). By integrating these parameters into the NorthWind Project's digital twin, predictive maintenance strategies can be refined, and situational awareness can be enhanced, reducing unplanned disruptions in critical communication links.

3. **Automatic Control and Optimization:** One of the key features of DTON is its ability to implement automatic control for optical networks. By integrating data from real-time monitoring and outputs from mirror modeling, the twin can autonomously configure network components, allocate resources efficiently, and optimize routing paths. Real-time insights enable dynamic routing adjustments to minimize latency and ensure data transmission continuity (Faruk and Savory, 2024). For offshore wind installations, this automation reduces manual interventions, streamlines energy distribution processes, and enhances the responsiveness of AI-driven systems.
4. **Enhancing Communication Reliability through GAN Integration:** As discussed earlier, the GANs framework can be adapted to the critical communication infrastructure assets as well. GANs can generate high-resolution simulations of network behavior under various stress conditions, such as extreme weather or physical damage to cables. By creating synthetic data for rare or extreme conditions, GANs facilitate comprehensive failure analysis, enabling the digital twin to predict and respond effectively to potential disruptions (Faruk and Savory, 2024). GAN-enhanced data improves the robustness of the digital twin's simulations, providing better insights into how communication infrastructure affects overall twin performance.

Figure 7.3 presents the extended framework model, expanding on the existing methodology that incorporates generative AI within digital twins for offshore wind farms. This enhanced framework integrates critical communication infrastructure, including optical network-based systems, into the asset layer and establishes seamless connections with the data transmission and reception layers. This design not only improves the functionality and reliability of the digital twin but also offers adaptability for application in other distributed renewable energy systems.

7.2.4 Results and Impact

The methodology of using generative AI to develop digital twins has shown impressive results, significantly improving the integration of AI within these models. This approach, especially evident in the offshore wind farm case study, effectively tackles the challenges and gaps presented in this chapter, leading to a substantial impact on renewable energy systems.

1. **Addressing Data Scarcity Using GANs:** One of the key outcomes of the NorthWind Project was the successful application of generative adversarial networks (GANs) to upscale low-resolution wind field simulations. Key results include:

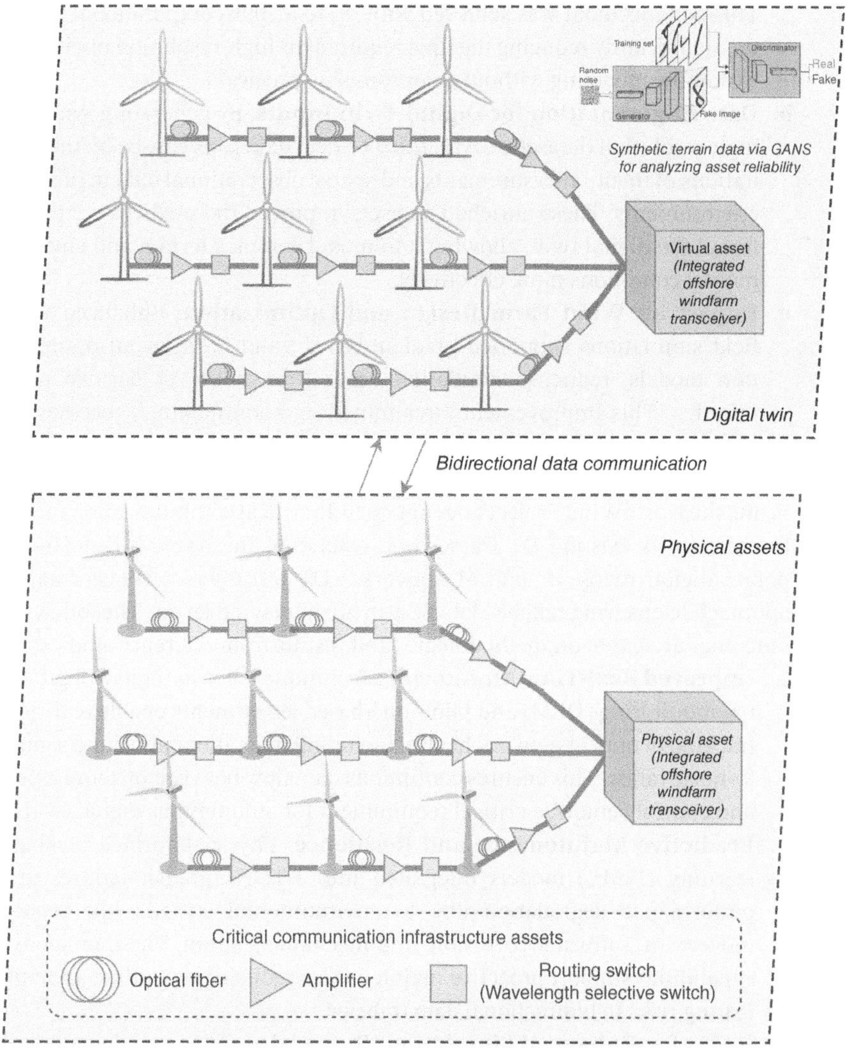

Figure 7.3 Enhanced digital twin framework illustrating the integration of critical communication infrastructure—comprising optical fibers, amplifiers, and wavelength-selective routing switches. The virtual layer leverages synthetic terrain data generated through GANs. *Source*: Hanu Indiran.

a. **High-Resolution Wind Field Simulations:** GANs enabled the transformation of coarse-resolution wind flow data (800m × 800m) into high-resolution outputs (200m × 200m), achieving an accuracy comparable to trilinear interpolation methods used in traditional modeling approaches.

This enhancement was achieved with up to a 100x computational speed-up, significantly reducing the time required for high-resolution microscale wind flow modeling without compromising precision.

b. **Data Augmentation for Digital Twin Inputs:** By generating synthetic high-resolution datasets, GANs filled critical data gaps caused by the limitations of in-situ measurements and sparse observational data in offshore environments. These enriched datasets improved the predictive capabilities of the digital twin, allowing it to model complex terrain and environmental conditions more effectively.

c. **Impact on Wind Farm Design and Optimization:** Enhanced wind field simulations supported gradient-based wind farm layout optimization models, reducing simulation times from hours or days to mere minutes. This improvement streamlined decision-making processes for turbine placement and energy yield maximization.

2. **Addressing Challenges in Critical Communication Infrastructure:** While the NorthWind Project does not include critical communication infrastructure in its existing DT framework, extending the framework to incorporate digital twins for optical networks (DTON) offers a transformative approach to ensuring reliable data flow in offshore wind farms. The following outcomes are based on methodologies and results from reference studies:

   a. **Improved Real-Time Monitoring:** Techniques such as digital longitudinal monitoring (DLM) and pilot-tone-based assessments enable real-time tracking of optical network health, including fiber attenuation and signal-to-noise ratios. This ensures continuous data flow between offshore assets and control centers, a critical requirement for autonomous digital twins.

   b. **Predictive Maintenance and Resilience:** Physics-informed machine learning (PIML) models integrated into DTONs predict failures and performance degradation due to environmental stresses like seabed movement, saltwater corrosion, and mechanical strain. These predictive capabilities support proactive maintenance, reducing downtime and mitigating risks to bidirectional data transfer.

   c. **GANs for Enhanced Simulation:** By applying GANs to generate synthetic high-resolution datasets for environmental and operational conditions, DTONs can simulate communication network behavior under extreme scenarios. This capability is crucial for understanding the impact of communication reliability on digital twin efficiency.

   d. **Autonomous Network Optimization:** DTONs implement automated routing and resource allocation, ensuring seamless operation even during network disruptions. This enhances the digital twin's responsiveness and reliability in managing offshore wind farms.

Synthetic data generation through GANs ensures high-quality inputs for digital twins, reducing reliance on expensive and time-consuming in-situ measurements. Real-time monitoring and predictive modeling of optical networks ensure seamless data flow between digital and physical domains, enabling more autonomous and efficient operations.

7.3 Discussion

The transformative potential of Generative AI-driven digital twins, as demonstrated through the NorthWind Project, highlights a pivotal advancement in the integration of Artificial Intelligence into renewable energy systems. By addressing critical challenges such as data scarcity and communication infrastructure reliability, this methodology not only enhances the efficiency and scalability of digital twins but also sets a new benchmark for the future of renewable energy operations.

1. **Generative AI for Data Scarcity:** One of the most impactful aspects of this methodology is the application of Generative Adversarial Networks (GANs) to overcome data scarcity. The ability of GANs to generate high-resolution, realistic synthetic data from coarse inputs bridges critical gaps in atmospheric and environmental data for offshore wind farms. By achieving a 100x computational speed-up without sacrificing accuracy, this innovation has enabled real-time modeling and predictive analytics that were previously unattainable due to computational limitations. This success underscores the importance of combining advanced AI techniques with domain-specific modeling to address unique challenges in complex terrains.
2. **Innovative Framework Integration:** The integration of data acquisition, standardization, and computationally efficient modeling into the digital twin framework provided a comprehensive life-cycle management approach. The structured hierarchy of digital twin capabilities (ranging from standalone to autonomous operation) ensured systematic scalability, adaptability, and operational reliability. Additionally, the asset information model played a critical role in correlating diverse data streams, enabling seamless interaction between the physical and virtual layers of the system.
3. **Expanding to Critical Communication Infrastructure:** The proposal to extend the existing digital twin framework to include critical communication infrastructure, such as optical networks, demonstrates the versatility and scalability of the methodology. Techniques like digital longitudinal monitoring (DLM) and physics-informed machine learning (PIML) provide a roadmap for improving communication reliability and optimizing bidirectional data flow. The use of GANs for generating synthetic datasets further reinforces this

capability, ensuring robust simulations under extreme environmental and operational conditions.

4. **Cross-disciplinary Collaboration:** A significant contributor to the success of the NorthWind Project was the active collaboration between academia, industry, and public sector stakeholders. Academic institutions provided the foundational research and expertise in generative AI, digital twin development, and modeling techniques. Industry partners brought practical insights into offshore wind farm operations, ensuring that the solutions addressed real-world challenges. Public sector support, particularly from the Research Council of Norway, ensured sustained funding, alignment with policy goals, and stakeholder engagement. This synergy of expertise, resources, and objectives was instrumental in achieving the project's outcomes.
5. **Innovative Public Engagement:** The NorthWind Project's emphasis on transparent communication and informed decision-making ensured stakeholders' trust and alignment. By presenting intuitive and accessible results to diverse audiences, including policymakers, industry leaders, and the public, the project fostered a collaborative environment conducive to the adoption and scaling of the proposed solutions.

The methodologies demonstrated in the NorthWind Project are highly transferable, offering a scalable blueprint for other distributed renewable energy systems.

Generalization to Diverse Renewable Energy Systems

While the digital twin framework discussed has been illustrated with offshore wind energy, its foundational methodologies, particularly generative AI, are widely generalizable. For instance, in solar power systems, generative AI can synthesize high-resolution solar irradiance data to enhance forecasting and manage variability. Similarly, for hydropower plants, generative models can produce synthetic inflow datasets, supporting accurate predictions and adaptive reservoir operations. Further, in more complex hybrid renewable energy systems (e.g. wind–solar–storage hybrids), the framework could simulate and optimize interactions between diverse energy sources, ensuring reliable, scalable, and adaptive energy management across intermittent conditions.

Real-time Optimization Under Dynamic Conditions

Renewable energy sources inherently exhibit intermittent and unpredictable generation patterns. To achieve effective real-time optimization, the proposed digital twin framework continuously integrates real-time sensor inputs with synthetic data generated through generative AI, facilitating immediate and accurate modeling of rapidly changing conditions. Techniques such as Bayesian AI methods, which provide probabilistic modeling and dynamic uncertainty quantification, can further enhance this optimization. By continuously updating probabilistic predictions of energy generation, demand fluctuations, and potential infrastructure failures, Bayesian inference allows digital twins to dynamically adapt control and operational decisions, ensuring efficient management despite energy intermittency and operational uncertainty.

Enhancing Trust Through Explainable and Bayesian AI
Developing trust and confidence in digital twin models is critical for industrial adoption. Integrating explainable AI (XAI) methodologies, including causal inference techniques, helps stakeholders clearly identify relationships between inputs and AI-generated outcomes (Dwivedi et al., 2023). Moreover, Bayesian AI further enhances transparency by explicitly quantifying uncertainty in predictions, thereby providing stakeholders with probabilistic insights into model performance and reliability. By clearly communicating model confidence levels and prediction uncertainties, Bayesian methods significantly contribute to informed decision-making (Korb and Nicholson, 2010). Together, these explainable approaches—causal inference, Bayesian methods, and model interpretability techniques like SHAP—support transparency (Dwivedi et al., 2023), regulatory compliance, and acceptance of AI-driven optimization strategies in renewable energy sectors.

7.4 Conclusion and Future Work

The integration of Generative AI-driven digital twins into offshore wind farm operations, as explored in this chapter, represents a significant leap forward in addressing key challenges of data scarcity and communication reliability within renewable energy systems. By demonstrating the use of advanced AI methodologies such as generative adversarial networks (GANs) and proposing an extension of the digital twin framework to include critical communication infrastructure, this work lays the foundation for enhanced operational efficiency, scalability, and resilience in distributed renewable energy systems.

The achievements of this approach are manifold. The deployment of GANs to upscale atmospheric flow data has redefined the data generation process, enabling real-time, high-resolution modeling with unprecedented computational efficiency. This innovation not only bridges existing data gaps but also enhances the predictive and prescriptive capabilities of digital twins, optimizing wind farm design, layout, and operation. Moreover, extending the digital twin framework to include optical communication networks offers transformative potential for addressing the reliability and scalability challenges of critical communication infrastructure. Techniques such as real-time monitoring, predictive analytics, and automated control embedded within the digital twin for optical networks (DTON) framework pave the way for seamless bidirectional data flow and autonomous decision-making.

However, significant challenges remain that require continued exploration. One key challenge is the scalability of digital twin frameworks to accommodate the growing complexity and size of renewable energy systems. As offshore wind farms expand and integrate with hybrid renewable energy grids, the computational demands of real-time simulations and predictive analytics will increase exponentially. Addressing this will require advancements in reduced-order modeling, optimization algorithms, and edge computing capabilities to maintain real-time performance without compromising accuracy.

Another pressing issue lies in the environmental and ethical implications of deploying large-scale AI models. The energy-intensive nature of training and running sophisticated generative models, such as GANs, poses challenges to the sustainability goals that these technologies aim to advance. Future work must focus on developing energy-efficient AI architectures and exploring distributed training approaches to mitigate the carbon footprint of these solutions.

In future work, the integration of knowledge graphs presents considerable potential for organizing and interpreting the intricate relationships between data and models in renewable energy systems. The utilization of domain-specific ontologies in renewable energy systems can significantly enhance the organization and interpretation of complex relationships between data and models. By leveraging standardized taxonomies for system components, these ontologies provide cognitive capabilities to digital twins, as demonstrated in research on cognitive digital twins within the manufacturing sector. Such capabilities enable more intelligent simulations and analysis, offering detailed insights into renewable energy systems and facilitating their adaptation across diverse operational contexts.

Furthermore, integrating critical communication infrastructure into digital twin frameworks presents opportunities for further refinement and innovation. Establishing efficient protocols for seamless data acquisition, storage, and transfer across diverse systems and platforms is a promising area for ongoing research. Ensuring the robustness of communication networks under varying environmental conditions and their harmonious integration with existing renewable energy systems will benefit from targeted pilot projects and collaborative field studies.

Last, fostering cross-sector collaboration remains crucial. Expanding the involvement of diverse stakeholders—ranging from policymakers to local communities—will ensure that these innovations are not only technologically robust but also socially and economically sustainable. Building frameworks for transparent data sharing, ethical AI deployment, and equitable access to technological advancements will further strengthen the impact of digital twin solutions in the renewable energy domain.

Final Thoughts: While the advancements in utilizing Generative AI-driven digital twins are truly transformative, we are just at the outset of this journey.

The next challenge lies in scaling these solutions, improving their efficiency, and ensuring their sustainability to fully harness their potential in advancing the global transition to renewable energy. The insights gained and methodologies developed through this work provide a solid foundation for future innovations, guiding us toward smarter, more resilient, and sustainable energy systems.

Bibliography

MA Ahmed and Y-C Kim. Protection schemes of EPON-based communication network for offshore wind power farms, In *2013 International Conference on ICT Convergence (ICTC)*, pp. 1135–1139. IEEE, 2013.

MA Ahmed, J-K Pan, M Song, and Y-C Kim. Communication network architectures based on Ethernet passive optical network for offshore wind power farms. *Applied Sciences,* 6(3):81, 2016.

Z Allal, HN Noura, O Salman, and K Chahine. Machine learning solutions for renewable energy systems: Applications, challenges, limitations, and future directions. *Journal of Environmental Management,* 354:120392, 2024.

M Attaran and BG Celik. Digital twin: Benefits, use cases, challenges, and opportunities. *Decision Analytics Journal,* 6:100165, 2023.

H Cao, C Tan, Z Gao, Y Xu, G Chen, P-A Heng, and SZ Li. A survey on generative diffusion model. *arXiv.org [Online]*, 2022.

A Creswell, T White, V Dumoulin, K Arulkumaran, B Sengupta, and AA Bharath. Generative adversarial networks: An overview. *IEEE Signal Processing Magazine,* 35:53–65, 2018.

S. Cruzes Optical networks automation overview: A survey. *Journal of Sensor Networks and Data Communications*, 3(1), 144–162, 2023.

M Dicorato, G Forte, M Pisani, and M Trovato. Guidelines for assessment of investment cost for offshore wind generation. *Renewable Energy,* 36:2043–2051, 2011.

R Dwivedi, D Dave, H Naik, S Singhal, R Omer, P Patel, B Qian, Z Wen, T Shah, G Morgan, et al. Explainable AI (XAI): Core ideas, techniques, and solutions. *ACM Computing Surveys,* 55(9):1–33, 2023.

M El-Sayed, PJ Ibrahim, and F Gunzer. Investigation of the precision regarding fiber fault location with a commercial optical time domain reflectometer. In *7th International Symposium on High-capacity Optical Networks and Enabling Technologies*. pp. 237–241. IEEE, Cairo, Egypt, 2010.

MS Faruk and SJ Savory. Measurement informed models and digital twins for optical fiber communication systems. *Journal of Lightwave Technology,* 42:1016–1030, 2024.

Global Wind Energy Council. Global offshore wind report 2024. Online, 2024. https://gwec.net/wp-content/uploads/2024/06/GOWR-2024_digital_final_v2.pdf

KB Korb and AE Nicholson. *Bayesian Artificial Intelligence*, CRC Press., 2010.

X Liu. Quality of optical channels in wireless SCADA for offshore wind farms. *IEEE Transactions on Smart Grid*, 3:225–232, 2012.

DJ Lucia, PS Beran, and WA Silva. Reduced-order modeling: New approaches for computational physics. *Progress in Aerospace Sciences*, 40:51–117, 2004.

Z Lv and S Xie. Artificial intelligence in the digital twins: State of the art, challenges, and future research topics. *Digital Twin*, 1:12, 2022.

NedZero. Connectivity in the North Sea: Towards the communications network of the future. Online, (n.d.). https://nedzero.nl/en/news/connectivity-in-the-north-sea-towards-the-communications-network-of-the-future

A Rasheed, F Stadtmann, E Fonn, M Tabib, V Tsiolakis, B Panjwani, KA Johannessen, T Kvamsdal, O San, JO Tande, et al. Digital twin for wind energy: Latest updates from the northwind project. In *Proceedings of the ASME 2024 43rd International Conference on Ocean, Offshore and Arctic Engineering, Volume 1 Offshore and Arctic Engineering*, p. V001T01A045. ASME, Singapore, 2024.

R Sahoo, SK Sahu, and P Shanmugam. Estimation of the channel characteristics of a vertically downward optical wireless communication link in realistic oceanic waters. *Optics & Laser Technology*, 116:144–154, 2019.

G Shoukat, G Michel, M Coughlan, A Malekjafarian, I Thusyanthan, C Desmond, and V Pakrashi. Generation of synthetic CPTs with access to limited geotechnical data for offshore sites. *Energies*, 16(9):3817, 2023.

J Smith, K Rovang, N Signorelli, and J Chatten. Offshore wind power: Using existing geo-data to derisk offshore wind energy development and infrastructure investment decisions. In *Proceedings of the Offshore Technology Conference Brasil*, p. OTC-32921-MS. Offshore Technology Conference, 2023.

W Song, Z Ding, B Mao, B Xu, H Gong, Y Zhang, and C Zhao. Distributed optical fiber sensing technology for wind turbine monitoring. In *2023 21st International Conference on Optical Communications and Networks (ICOCN)*, pp. 1–3. IEEE, 2023.

M Suzuki and Y Matsuo. A survey of multimodal deep generative models. *Advanced Robotics*, 36:261–278, 2022.

F Tao, H Zhang, and C Zhang. Advancements and challenges of digital twins in industry. *Nature Computational Science*, 4:169–177, 2024.

T Uhl. Reliability of communication systems used in offshore wind farms. *Journal of Telecommunications and Information Technology*, 4:61–66, 2022.

D Wang, Y Song, Y Zhang, X Jiang, J Dong, FN Khan, T Sasai, S Huang, APT Lau, M Tornatore, et al. Digital twin of optical networks: A review of recent advances and future trends. *Journal of Lightwave Technology*, 42:4233–4259, 2024.

S Whistler, IGC Ashton, AC Pillai, and GD Quartly. Enhancing offshore wind farm met-ocean data accessibility: A machine learning approach with satellite-derived wave measurements in the Celtic Sea. In *Proceedings of the ASME 2024 43rd International Conference on Ocean, Offshore and Arctic Engineering*, p. V007T09A042. ASME, 2024.

Xu, H Omitaomu, F Sabri, S Zlatanova, S Li, X Song. and Y Leveraging generative AI for urban digital twins: A scoping review on the autonomous generation of urban data, scenarios, designs, and 3D city models for smart city advancement. *Urban Informatics*, 3(1), 29, 2024.

L Yang, D Zhang, and GE Karniadakis. Physics-informed generative adversarial networks for stochastic differential equations. *SIAM Journal on Scientific Computing*, 42(1):A292–A317, 2020.

EY Zhang, AD Cheok, Z Pan, J Cai, and Y Yan. From turing to transformers: A comprehensive review and tutorial on the evolution and applications of generative transformer models. *Sci*, 5:46, 2023.

8

Vision Transformer-based O&M Model for Condition Monitoring of Solar Panels

Anu Mehra, Anubhav Gautam, and Kunal Jindal

Amity School of Engineering and Technology, Amity University, Noida, Uttar Pradesh, India

8.1 Challenges

Overconsumption of fossil fuels such as oil, coal, and natural gas is the cause of their accelerated depletion, and they may not last for more than a few decades. Global environmental challenges and a surging demand for energy have led mankind to explore new opportunities for utilization of renewable energy resources. Solar energy stands out as the most easily available and environmentally friendly of all the renewable energy sources available today. Sunlight reaching our planet is approximately 1.8×10^{11} MW, which is much greater than the world's total energy consumption at present (Parida et al., 2011).

In the past few decades, two key factors have driven the widespread use of photovoltaics (PVs), which are a noticeable increase in solar panel efficiency, and a significant decrease in solar panel costs.

8.1.1 Practical Issues

Although solar cells signify a rapidly expanding and feasible renewable energy source, the efficacy of photovoltaic systems is limited by considerable practical difficulties. These challenges include efficiency losses due to temperature fluctuations, shading effects, environmental impact, material degradation, and a need for effective energy storage solutions to manage sporadic sunlight.

As temperatures rise, the performance of photovoltaic cells tends to decline significantly due to a drop in the open-circuit voltage and a surge in the short-circuit

Renewable Energy Transition with Artificial Intelligence: Challenge-driven Solutions, First Edition. Nina Dethlefs and Joyjit Chatterjee. Published 2026 by John Wiley & Sons, Inc.

current. These factors together critically reduce the overall power output (Sun et al., 2022). Specifically, a temperature rise of 5 °C can lead to a 2% decrease in power output, while a 20 °C increase can result in a 10.4% power output reduction (Yadav et al., 2023). High temperatures lead to thermal decay, which reduces the power output of photovoltaics and shortens the lifespan of solar cells (Karaca and Yaver, 2024).

This thermal sensitivity is particularly pronounced in polycrystalline silicon solar cells, which are more sensitive to temperature changes than monocrystalline cells (Xu, 2024). The efficiency of PV modules can decline due to a multitude of reasons including humidity, radiation apart from temperature. These factors cause degradation such as corrosion, discoloration, and breakage, significantly compromising performance (Munoz et al., 2011; Kim et al., 2021).

Further, environmental factors such as partial or sharp shading can also lead to mismatch losses. While partial shading caused by moving clouds produces gentle irradiance transitions for large-scale PVs, about 70% of the total mismatch losses were caused by shadow edges with shading strengths ranging between 40% and 80% (Lappalainen and Valkealahti 2017; Maghami et al., 2016).

In large sovereign systems such as urban cities, there is a peak in energy demand during the night when the sun sets, compelling reliance on traditional energy sources (Popovych et al., 2023).

Dust and other air pollutants are yet another common cause for the lower efficiency of solar cells. The extent to which efficiency declines is influenced by the size of dust particles that settle on the surface of the PV modules. As the amount of dust increases, both the power output and the efficiency of the module tend to drop. In urban areas where smaller particles like cement are present, the power output diminishes even more since these tiny particles obstruct additional radiation on the PV module surface (Hussain et al., 2017; Duffie and Beckman, 1980; El-Shobokshy and Hussein, 1993). In places where air pollution is an issue, even a two-month outdoor exposure can cause up to 6.5% reduction in energy production (Sarver et al., 2013). During long periods without rain, daily energy losses can be higher than 20%. This reduction can be as high as approximately 40% in deserts. In general, $10 g/m^2$ accumulated dust on a solar PV module can reduce it's power by up to 25%. When $50 g/m^2$ of dust is spread uniformly, then power output is reduced by approximately 55–63% (Hussain et al. (2017)). These practical issues are summarized in Table 8.1.

8.1.2 Computer Vision Approaches in Photovoltaic

Maintenance and design of PV units, keeping their physical location in mind, can be expensive, and therefore, computer vision has been deployed to predict and help monitor their condition.

Table 8.1 Summary of practical issues with photovoltaics.

Key issues	Effects on performance	References
High temperatures	Short-circuit currents	Sun et al. (2022)
	Thermal decay reduced lifespan	Yadav et al. (2023)
		Karaca and Yaver (2024)
		Xu (2024)
Humidity	Corrosion, discoloration,	Munoz et al. (2011)
	and breakage	Kim et al. (2021)
Shading	Mismatch Losses	Lappalainen and Valkealahti (2017)
		Maghami et al. (2016)
Urban cities	Peak in energy demand at night	Popovych et al. (2023)
Dust particle	Reduced solar irradiance	Duffie and Beckman (1980)
	depending on particle	El-Shobokshy and Hussein (1993)
	size and air pollution	Sarver et al. (2013)
		Hussain et al. (2017)

8.1.2.1 Vision Transformers

Convolutional neural networks (CNNs) and recurrent neural networks (RNNs) have long been the standard for computer vision tasks. However, Vision Transformer models have recently emerged as a powerful alternative, often achieving comparable or even superior performance.

The concept of transformer was first introduced by Vaswani (2017). A Transformer is a neural network architecture that utilizes the self-attention mechanism, assigning varying levels of weightage to different components of the input data. Vaswani (2017) demonstrated that a Transformer can be trained faster than CNN- or RNN-based architectures for translation tasks. Unlike RNNs, Transformers can efficiently process non-sequential data. Transformers utilize attention mechanism and have the advantage that they can give preference to contextually relevant data. On the other hand, RNNs can access only immediately preceding state. Also, Transformers make use of positional embeddings to keep track of where each word is located in a sentence. In natural language-processing tasks, this helps the model to understand the order of words, which helps in comprehending the meaning of a given sentence.

The novel idea of Transformers was extended to Vision Transformer for computer vision tasks by Dosovitskiy (2020). Vision transformers take 2D images as input. The input image is divided into sequential patches. These patches are converted into a single vector by merging the channels of all pixels within that patch and subsequently linearly mapping the vector to the specified input dimension. Positional embedding is added to the patches, which allows the model to focus on desired areas of the image. The input sequence is fed to a standard transformer. The model is then pretrained with image labels, and further improved on subsequent dataset for identifying and classifying images. The architecture used in our work based on Vision Transformer is shown in Figure 8.1. Figure 8.2 shows the graphical representation of the Transformer Encoder shown in Figure 8.1.

Some studies that have implemented Vision Transformers in the field of photovoltaics are highlighted here. Guo et al. (2024) have improvised the use of FCNs, which had limited localized mechanisms in convolution operations by the use of Vision Transformers. Mercier et al. (2024) have used Vision Transformers to measure solar irradiance. Barnabé et al. (2023) used Vision Transformers to quantify damages and detect flaws in monocrystalline PV cells. This chapter proposes the use of Vision Transformer's base model, which has approximately 86M parameters only. This makes it convenient to run the module not only on cloud environments but also locally on our personal devices such as mobile phones, while providing reasonable accuracy, as discussed in the following section. Further, we have also proposed a way to integrate LLMs along with Vision Transformers. The predictions from Vision Transformer are used as context for the LLM to predict O&M actions for the different fault types, thus acting as an interactive decision support system.

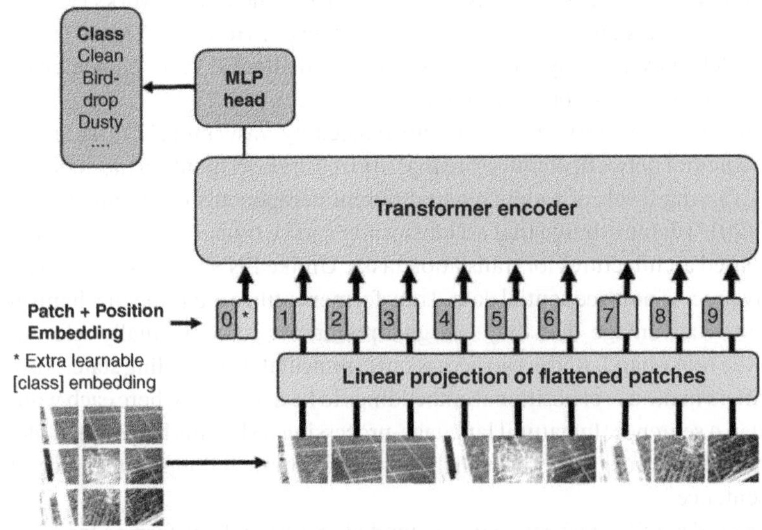

Figure 8.1 Model overview adapted from Dosovitskiy (2020). The image is split into fixed-size patches, which are linearly embedded and augmented with positional information to preserve spatial context. The resulting sequence of vectors is fed to a standard Transformer encoder. An extra learnable "classification token" is added to the sequence to perform classification.

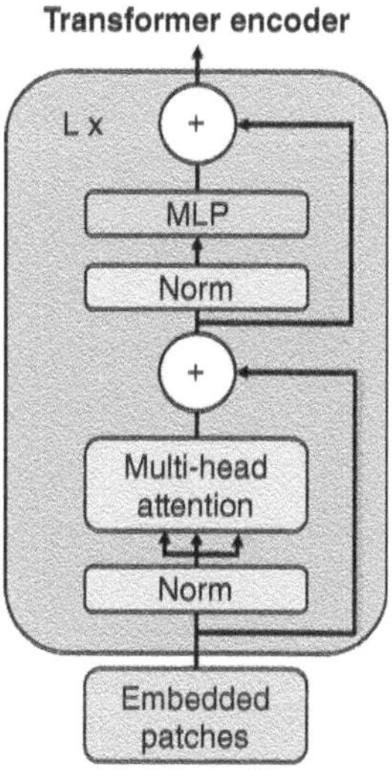

Figure 8.2 Graphical representation of transformer encoder shown in Figure 8.1.

8.1.3 Using Large Language Models for Interpreting Vision Transformer Results

Vision Transformer results have been effectively applied to image classification tasks. However, a challenge arises in interpreting their numerical outputs, such as class labels and confidence scores. To address this, the integration of Large Language Models (LLMs) with Vision Transformers is proposed to translate these outputs into meaningful, context-rich information.

Dai et al. (2022) proposed how LLMs can enhance imaging tasks by combining vision and language models. Wang et al. (2024) show how LLMs can enhance vision-based tasks by combining vision and language models. Applying this approach to PV panel classification can enable more interpretable and automated inspection processes in photovoltaics as well.

8.2 Case Study

This case study explores the application of a hybrid vision-language model for fault detection in PV cells. By using Vision Transformers (ViT), we have classified the input images of PV cells into common fault types. To enhance the diagnostic process beyond simple classification, an LLM was integrated with the Vision Transformer. The output of the Vision Transformer's linear classification layer, representing the predicted fault class, is fed as input to the LLM. Figure 8.3 shows

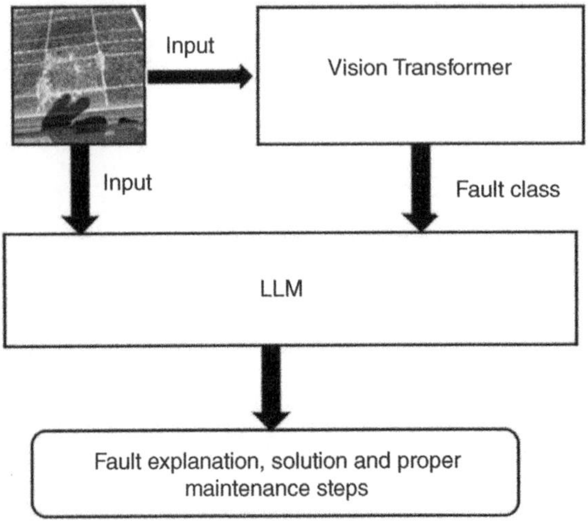

Figure 8.3 The proposed framework integrates a Vision Transformer to classify solar panel faults from image inputs. The fault class is passed to a Large Language Model (LLM), which generates detailed fault explanations, suggested solutions, and step-by-step maintenance strategies.

the workflow of this project. This enables the generation of context-rich interpretations, providing detailed descriptions of identified faults, their potential causes, and their impact on cell performance. This combined approach not only automates fault detection but also offers valuable insights for maintenance and preventive measures, bridging the gap between computer vision-based diagnostics and actionable expert knowledge.

8.2.1 Model Description

A transfer learning approach was employed to create the classification model. Transfer learning involves taking a model that is already trained on a large dataset and fine-tuning it on a smaller dataset. The dataset used in this study was obtained from Kaggle.[1] It contains 869 solar panel images belonging to six different classes, namely, Clean, Dusty, Bird-drop, Electrical-damage, Physical-damage, and Snow-covered. An 80-20 split was applied to the dataset, resulting in 695 training images and 174 testing images. Table 8.2 contains a category-wise distribution of training data, which has 695 images, and Table 8.3 contains a category-wise distribution of testing data, which has 174 images. Sample images from the six predictable classes are shown in Figure 8.4.

1 Dataset URL: https://www.kaggle.com/datasets/pythonafroz/solar-panel-images

Table 8.2 Training data consisted of 695 images.

Class	Number of images
Bird-drop	147
Clean	152
Dusty	152
Electrical-damage	92
Physical-damage	54
Snow covered	98

Table 8.3 Test data consisted of 174 images.

Class	Number of images
Bird-drop	44
Clean	41
Dusty	38
Electrical-damage	11
Physical-damage	15
Snow covered	25

Figure 8.4 Sample images of all classes in our dataset.

In order to prevent overfitting and to enable the model to be more adaptable to real-world scenarios, the data are initially augmented. Data augmentation consists of applying random transformations to training images to enhance the diversity of the dataset artificially. This contributes to enhancing the model's resilience and its ability to generalize to unfamiliar data. The images were randomly cropped, and framing variability was introduced for better generalization. The images were then flipped horizontally with a probability of 50% to account for different orientations. Color Jitter was then introduced to account for varying lighting conditions as per Table 8.4.

8.2.2 Vision Transformer Applied to Photovoltaic Cells

Table 8.5 describes the Vision Transformer Architecture used. We have fine-tuned the Google ViT-base-patch16-224-in21k Vision Transformer model, which is a base-sized model pretrained on the ImageNet-21K dataset (14 million images, 21 843 classes). The model takes an image of 224 × 224 resolution as input and splits it into 16 × 16 patches, while also adding linear embedding to them. Then, the patches are treated as tokens like words in Natural Language Processing, and are processed through the self-attention mechanism of the transformer model.

Table 8.4 Data augmentation.

Augmentation	Parameters
Random crop	224 × 224
Random horizontal flip	Probability: 50%
Color Jitter	Brightness ±0.2, Contrast: ±0.2
	Saturation: ±0.2, Hue: ±0.1

Table 8.5 Vision transformer architecture.

Parameter	Dimension
Input Size	224 × 224
Patch Size	16 × 16
Linear Embedding Dimensions	768
Number of Layers	12
Number of Hidden Units	768
Feedforward Layer Dimension	3072
Output Size	6

8.2.3 Vision Transformer Comparison with Other Models

The Test Loss, Test accuracy, and f1-score of vit-base-patch16-224-in21k are compared with VGG-16, VGG-19, ResNet18, ResNet50, MobileNetV2, MobileNetV3 (Small), and MobileNetV3 (Large) in Table 8.6. Our key results were as follows:

- **Lowest Test Loss:** ViT achieved the lowest test loss of 0.3305 among all evaluated models. This indicates that ViT generalizes better to unseen data compared to other architectures.
- **Highest Test Accuracy:** ViT records a test accuracy of 94.83%, significantly higher than the other models. This demonstrates that the ViT is significantly more likely to classify images correctly compared to the other evaluated models.
- **High F1-score:** ViT achieved an F1-score of 0.95, outperforming all other models. The F1-score, which balances precision and recall, is particularly critical for imbalanced datasets like this one, where certain categories are less represented. ViT's high F1-score demonstrates its effectiveness in correctly classifying both common and rare classes of the dataset.

8.2.4 Attention Map Images

Attention maps in ViTs are essential for understanding how the model interprets visual data, as they highlight the regions of an image that the model focuses on for making predictions. In the context of solar panel image analysis, these maps help determine whether the model is prioritizing critical features like cracks, dust, or discolorations over irrelevant background elements such as the sky or ground. Brighter areas in the attention maps indicate higher attention, meaning the model assigns greater importance to those regions. This is particularly important in solar

Table 8.6 Model performance comparisons.

Model architecture	Test loss	Test accuracy	F1-score
ViT-base-patch16-224-in21k	0.3305	0.9483	0.95
VGG-16	0.5566	0.8448	0.84
VGG-19	0.9209	0.8161	0.82
Resnet18	0.5567	0.8391	0.84
Resnet50	0.5003	0.8908	0.89
MobileNetV2	0.5154	0.8621	0.86
MobileNetV3 (Small)	0.5259	0.8621	0.86
MobileNetV3 (Large)	0.5802	0.8851	0.88

panel images, where the panel sometimes occupies only a small portion of the image, and the model must learn to ignore distractions to accurately detect faults. Fine-tuning the model helps refine these attention patterns, ensuring it focuses on relevant areas for improved accuracy.

The attention maps of the pretrained ViT show a more generalized focus across the image, as it has not yet specialized in detecting features or faults in solar panels. Therefore, it remains uncertain where precisely the model's attention is directed in the image. This might lead to the model prioritizing irrelevant areas in certain images where the solar panel occupies a fraction of the image, such as the background areas—the sky, the ground, or surrounding objects—thus ignoring the relevant areas like cracks, accumulated dust, and discolorations.

After fine-tuning, the ViT attention maps specifically focus on the critical areas of the solar panel. In layers 1–4, significant localization of attention is seen as compared to the pre-fine-tuning maps. The area around the bird-drop and the edges is brighter than the other regions as seen in Figure 8.7, indicating high attention scores assigned to that part of the image.

In layers 5–8, uniformity in the attention scores is assigned to all parts of the image, except the edges of the solar cells and the bird-drop region, which is better than the irregularities in the pre-fine-tuning maps. In the last four layers, increased attention is given to individual solar cells rather than the edges as seen in the previous layers. Although less significant than the previous layers, the attention does localize a bit around the bird-drop region. Attention maps are prepared for the image of Bird Drop on a solar cell as shown in Figure 8.5. The comparison of attention maps before and after fine-tuning can be seen in Figures 8.6 and 8.7, respectively.

Figure 8.5 Image of class "Bird-drop" for which attention maps are prepared.

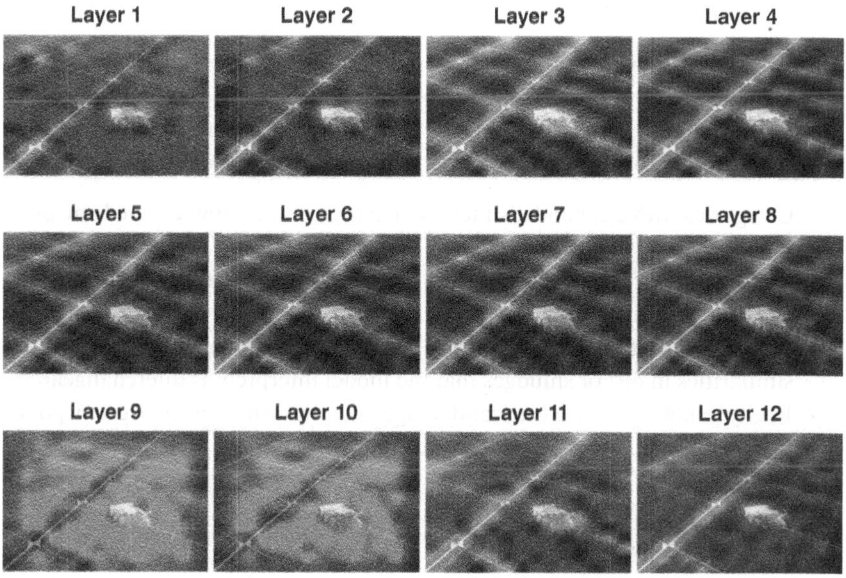

Figure 8.6 Attention maps before fine-tuning on PV panel image dataset.

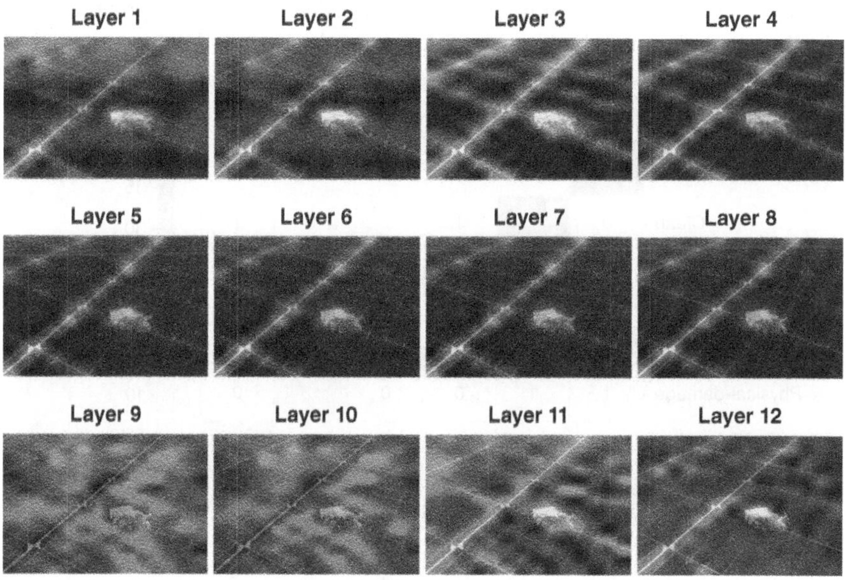

Figure 8.7 Attention maps after fine-tuning on PV panel image dataset.

8.2.5 Confusion Matrices

The confusion matrix for Vision Transformer Model is shown in Figure 8.8. Confusion matrices for ResNet50 and ResNet18 models are given for reference in Figures 8.9 and 8.10, respectively. The following conclusions can be drawn regarding the overall performance of Vision Transformer Model from Figure 8.8.

- Categories with clear and distinct visual patterns (e.g. snow-covered and bird-drop) were classified accurately.
- Overlapping or subtle features like electrical-damage and physical-damage are more challenging for the model to classify accurately.
- Dusty and clean panels show some confusion, which could be due to visual similarities in dirt or smudges that the model interprets as interchangeable.
- Faulty panels such as electrical-damage and physical-damage overlap, possibly due to shared visual features like cracks or other abnormalities.

8.2.6 Model Fine-tuning Process

In this study, we implemented a fine-tuning approach utilizing the ViT architecture, specifically the base model pretrained on ImageNet-21k, for solar panel defect classification. The fine-tuning procedure included a standard preprocessing pipeline that featured both normalization methods and data augmentation

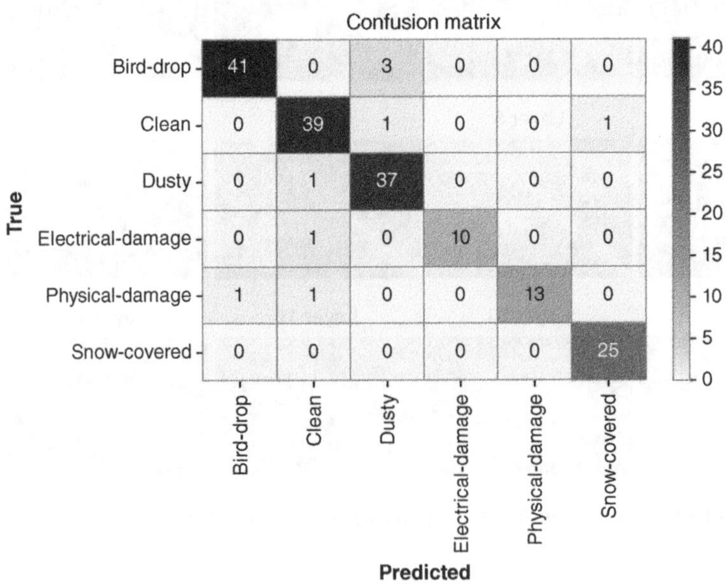

Figure 8.8 Confusion matrix generated for Vision Transformer (ViT) model.

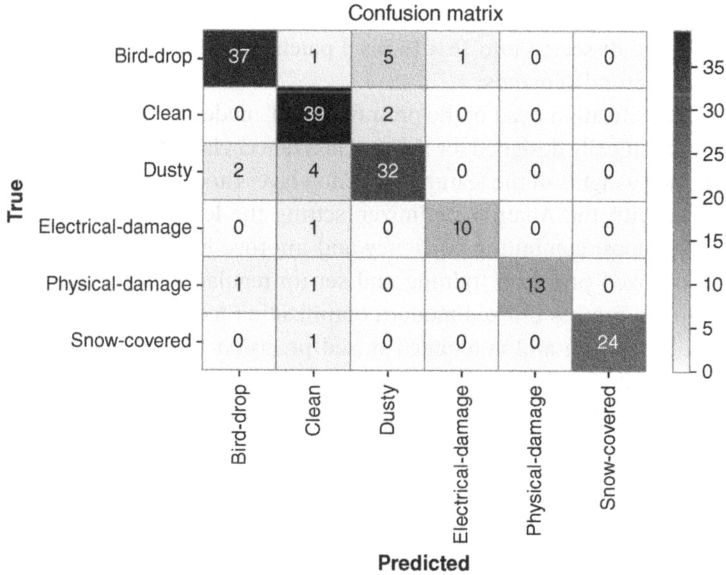

Figure 8.9 Confusion matrix generated for ResNet50 model.

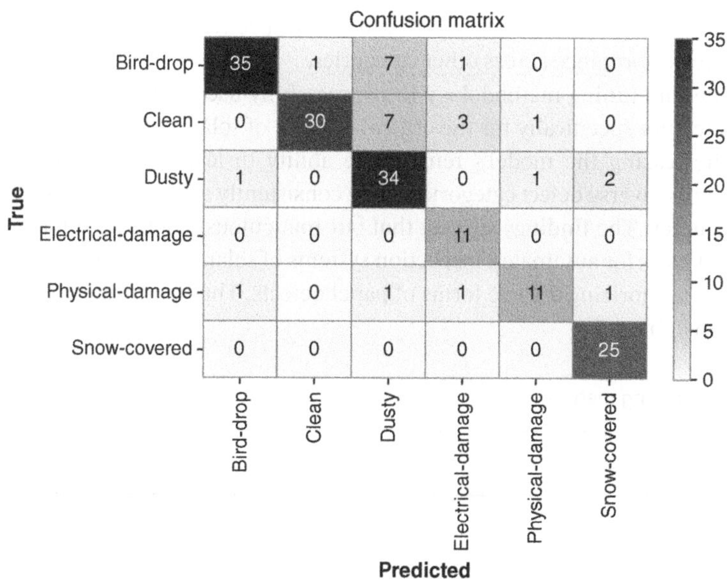

Figure 8.10 Confusion matrix generated for ResNet18 model.

approach. Throughout the training phase, images were vigorously resized to 224 × 224 pixels and dissected into 16 × 16 pixel patches, perfectly aligning with the model's architectural demands.

The original classification head of the pretrained ViT model was replaced with a custom layer specifically designed for the six-class defect classification task. We kept the pretrained weights in the feature extraction layers and used a smart fine-tuning approach with the AdamW optimizer, setting the learning rate to 2e-4 for 15 epochs. To boost computing efficiency and improve how well the model works, we used mixed precision training and set up regular evaluation checkpoints. The training process utilized modern optimization techniques, including dynamic batch processing and automated mixed precision training. The training process consisted of recent optimization methods, featuring dynamic batch processing and automated mixed precision training. The model's effectiveness was consistently assessed via validation phases, with the top-performing model checkpoint being retained according to validation criteria. This approach ensured optimal model selection while preventing overfitting. The fine-tuning process demonstrated remarkable effectiveness, achieving a validation accuracy of 92.7% across all defect categories. The model showed remarkable performance in identifying specific defect types, with F1-scores ranging from 86% to 98% across different categories. The electrical damage and snow-covered defects were particularly well-classified, achieving F1-scores of 97% and 98% respectively, while maintaining reasonable performance across other categories.

This unique fine-tuning methodology has successfully used the Vision Transformer architecture specifically for the critical domain of solar panel defect classification, showcasing the model's remarkable ability to learn and generalize effectively across diverse defect categories while consistently achieving good classification accuracy. The findings suggest that our meticulously optimized model offers a robust basis for automated inspection systems of solar panels, proficiently detecting and categorizing diverse forms of panel defects. The fine-tuning metric is described in Table 8.7.

Table 8.7 Fine-tuning metrics.

Metric	Value
Epochs	25
Learning rate	1e-04
Batch size (training)	16
Batch size (validation)	8
Optimizer	Adam with betas = (0.9, 0.999) and epsilon = 1e-08

8.3 Future Work

Building on the findings of this study, future research could explore the following directions:

- **Real-world Deployment:** Deploying these models in real-world scenarios will assess the practicality of the proposed system. Creating a proper pipeline/workflow for its application is important for making it usable for end users.
- **Exploration of Larger Dataset:** Expanding the study to include larger and more diverse datasets to validate the scalability and robustness of the models.
- **Explainability using LLMs:** Developing visual and textual interpretability frameworks to demystify the decisions made by the integrated system. And using fine-tuned LLMs to reduce the dependency on third party such as OpenAI will reduce the cost and foster greater trust among end-users.
- **User-friendly Query Systems:** Building interactive systems where users can query images or datasets using natural language, thus making it easier for users to take benefit of these complex systems.
- **Data Augmentation and Preprocessing:** Implement advanced augmentation strategies to improve model robustness, especially in imbalanced or noisy datasets.

For the sake of simplicity and to demonstrate the idea, which can be expanded on in the future, we show an example by using ChatGPT (GPT-4o) as our LLM. In practice, it would be integral to fine-tune any LLM with domain-specific PV industry data such as O&M work orders, past maintenance action records, etc., for generating more robust O&M strategies. Figure 8.11 illustrates how the outcomes from a Vision Transformer can be utilized as inputs for the LLM. This provides context for the outcomes and allows individuals with limited knowledge to easily understand the results. Through prompt engineering, we have generated a precise explanation of the fault and outlined the steps to resolve it along with maintenance advice to avert similar problems in the future.

8.4 Conclusion

This study evaluated the performance of various deep learning models for the task of PV panel anomaly classification, highlighting the model comparisons. The results demonstrate that the ViT-based model achieved superior accuracy (94.83%) with the lowest test loss (0.3305), outperforming other deep learning models like ResNet, VGG, and MobileNet variants. This indicates that transformer-based architectures are particularly effective for the given dataset. While other models,

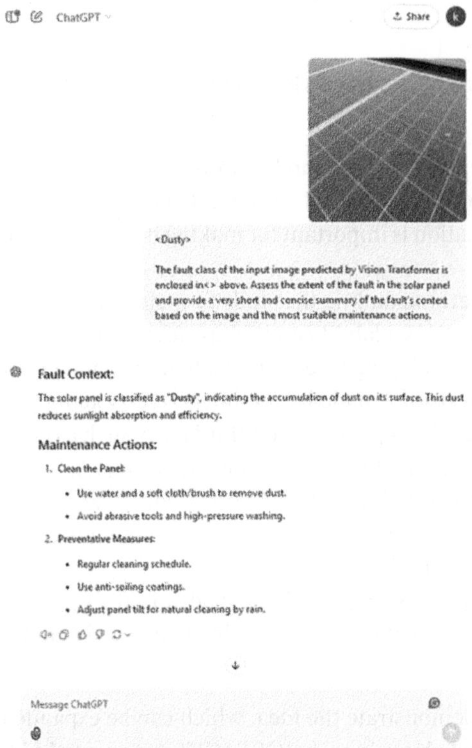

Figure 8.11 Example depiction of integrating LLM with the proposed fault prediction model. Generated with AI using-OpenAI.

such as ResNet50 and MobileNetV3 (Large), also exhibited competitive performance, their slightly higher test loss and lower accuracy underline the trade-offs between model complexity and generalization capabilities.

A key contribution of this study is the integration of Vision Transformer outputs with ChatGPT, which is an advanced Large Language Model. ChatGPT was used to clarify the irregularities discovered by the model and produce detailed, user-friendly summaries and operational maintenance advice for the user by analyzing the ViT results. This integration offers a deeper insight into identified anomalies, rendering the outcomes understandable and easy to implement for stakeholders who do not possess technical expertise. This hybrid system showcases the potential of combining vision-based AI with natural language processing (NLP) to enhance interpretation and usage, particularly in solar panel fault diagnostics.

This study establishes a foundation for context-aware solutions that integrate vision-based AI and LLMs, bridging the gap between technical insights and practical applications. Such efforts will help extend the impact of Artificial Intelligence across diverse domains, and help provide solutions to mankind diverse issues and challenges.

Bibliography

João Pedro Costa Barnabé, Lenin Patricio Jiménez, Gustavo Fraidenraich, Eduardo Rodrigues De Lima, and Harrison Franca Dos Santos. Quantification of damages and classification of flaws in mono-crystalline photovoltaic cells through the application of vision transformers. *IEEE Access*, 11:112334–112347, 2023.

Damai Dai, Yutao Sun, Li Dong, Yaru Hao, Shuming Ma, Zhifang Sui, and Furu Wei. Why can GPT learn in-context? Language models implicitly perform gradient descent as meta-optimizers. *arXiv preprint arXiv:2212.10559*, 2022.

Alexey Dosovitskiy. An image is worth 16x16 words: Transformers for image recognition at scale. *arXiv preprint arXiv:2010.11929*, 2020.

John A Duffie and William A Beckman. *Solar Engineering of Thermal Processes*. Wiley, New York, 1980.

Mohammad S El-Shobokshy and Fahmy M Hussein. Effect of dust with different physical properties on the performance of photovoltaic cells. *Solar Energy*, 51(6):505–511, 1993.

Zhiling Guo, Jiayue Lu, Qi Chen, Zhengguang Liu, Chenchen Song, Hongjun Tan, Haoran Zhang, and Jinyue Yan. Transpv: Refining photovoltaic panel detection accuracy through a vision transformer-based deep learning model. *Applied Energy*, 355:122282, 2024.

Athar Hussain, Ankit Batra, and Rupendra Pachauri. An experimental study on effect of dust on power loss in solar photovoltaic module. *Renewables: Wind, Water, and Solar*, 4:1–13, 2017.

Cengiz Karaca and Seren Yaver. Determining the effect of photovoltaic module surface temperature on generation efficiency. *Revista Ingenieria E Investigacion*, 44(2):e106383, 2024. https://doi.org/10.15446/ing.investig.106383.

Jaeun Kim, Matheus Rabelo, Siva Parvathi Padi, Hasnain Yousuf, Eun-Chel Cho, and Junsin Yi. A review of the degradation of photovoltaic modules for life expectancy. *Energies*, 14(14):4278, 2021.

Kari Lappalainen and Seppo Valkealahti. Photovoltaic mismatch losses caused by moving clouds. *Solar Energy*, 158:455–461, 2017.

Mohammad Reza Maghami, Hashim Hizam, Chandima Gomes, Mohd Amran Radzi, Mohammad Ismael Rezadad, and Shahrooz Hajighorbani. Power loss due to soiling on solar panel: A review. *Renewable and Sustainable Energy Reviews*, 59:1307–1316, 2016.

Thomas M Mercier, Amin Sabet, and Tasmiat Rahman. Vision transformer models to measure solar irradiance using sky images in temperate climates. *Applied Energy*, 362:122967, 2024.

MA Munoz, M Carmen Alonso-García, Nieves Vela, and Faustino Chenlo. Early degradation of silicon PV modules and guaranty conditions. *Solar Energy*, 85(9):2264–2274, 2011.

Bhubaneswari Parida, Selvarasan Iniyan, and Ranko Goic. A review of solar photovoltaic technologies. *Renewable and Sustainable Energy Reviews*, 15(3):1625–1636, 2011.

Vitalii Popovych, Yevhen Rudniev, Rostyslav Brozhko, Vadym Tarasov, Oleksandr Tymchenko, and Ihor Aliekperov. Photovoltaics: Problems, challenges, environmental feasibility and efficiency. In *2023 IEEE 5th International Conference on Modern Electrical and Energy System (MEES)*, pp. 1–6. IEEE, 2023.

Travis Sarver, Ali Al-Qaraghuli, and Lawrence L Kazmerski. A comprehensive review of the impact of dust on the use of solar energy: History, investigations, results, literature, and mitigation approaches. *Renewable and Sustainable Energy Reviews*, 22:698–733, 2013.

Chunlei Sun, Yuan Zou, Caiyan Qin, Bin Zhang, and Xiaohu Wu. Temperature effect of photovoltaic cells: A review. *Advanced Composites and Hybrid Materials*, 5(4):2675–2699, 2022.

A Vaswani. Attention is all you need. *Advances in Neural Information Processing Systems*, 2017.

Wenhai Wang, Zhe Chen, Xiaokang Chen, Jiannan Wu, Xizhou Zhu, Gang Zeng, Ping Luo, Tong Lu, Jie Zhou, Yu Qiao, et al. Visionllm: Large language model is also an open-ended decoder for vision-centric tasks. *Advances in Neural Information Processing Systems*, 36, 2024.

Wenxi Xu. How important of the effect of temperature on the efficiency of solar photovoltaic cells? *Advances in Engineering Innovation*, 10(1):85–100, 2024.

Apurv Yadav, Walid Ayadi, and Mohammad Khalid. The effect of temperature on photovoltaic power generation. In *2023 International Conference on Computational Intelligence and Knowledge Economy (ICCIKE)*, pp. 473–477. IEEE, 2023.

9

Artificial Intelligence Applications: Case Studies from Challenging Domains

Saumya Bhatnagar-Knox,[1] Sandra Castellanos-Paez,[1] Rongge Guo,[2] Francesco Percassi,[1] Megan Robinson,[1] and Mauro Vallati[1]*

[1] *School of Computing and Engineering, University of Huddersfield, Queensgate, Huddersfield, UK*
[2] *School of Traffic and Transportation, Beijing Jiaotong University, Beijing, China*

*Corresponding author: Mauro Vallati; m.vallati@hud.ac.uk

9.1 Introduction

Artificial Intelligence approaches are increasingly being used in a wide range of real-world applications. In this chapter, we present three case studies where artificial intelligence (AI) has been exploited in challenging domains. Those domains are characterized by the complexity of the problems addressed, and by the shared potential for positive societal impact.

The first case study, which will be described and analyzed in detail, explores the use of AI in urban mobility and traffic control, where AI has been used to reduce traffic congestion and contribute to more sustainable and efficient public transport systems. The second case study examines the use of machine learning approaches for improving postconsumer textile sorting. While the presented work is still in its infancy, the discussion highlights the challenges and the expected benefits in terms of improved waste recycling. Finally, the third case study investigates the role of AI in managing energy distribution networks more efficiently. AI-driven systems can optimize energy consumption, integrate renewable energy sources, and improve grid reliability, leading to reduced carbon emissions and enhanced energy security.

By examining these case studies, we aim to highlight the innovative applications of AI, their measurable contributions to environmental preservation, and their potential for widespread adoption and adaptation. To support the use of AI in different domains, we take the occasion of presenting the case studies to discuss some lessons learnt and the challenges faced in the design process.

9.2 Urban Traffic Control

Due to the growing urbanization trend and the growing population, urban infrastructure is being pushed to its limit. Traffic congestion has emerged as a major challenge for many urban areas, especially during peak hours, significantly hindering the efficiency of urban mobility. This issue not only causes delays and inconvenience for travelers, but also leads to a range of negative consequences. First, it contributes to higher levels of pollution when vehicles are stuck in traffic, which also adversely affects public health: transportation as a sector contributes to approximately 25% of worldwide greenhouse gas emissions. Additionally, traffic congestion slows down the flow of goods and services, thereby impeding economic productivity (Levy et al., 2010). These effects place a heavy burden on both the environment and the economy, making it necessary to address congestion through innovative urban planning and transportation solutions. In this section, we describe three areas of urban mobility where AI techniques have been successfully applied: traffic signal optimization, traffic distribution, and public transport improvements.

Integrating AI techniques, particularly those based on automated planning, into urban traffic control (UTC) has gained growing interest to smooth traffic flows through traffic signal optimization. Traffic signal optimization is the process of defining the timing of traffic light phases at an intersection to reduce overall delay for road users. This includes determining the duration of each phase, the overall cycle length, and the fixed intergreen periods between phases. These parameters are subject to constraints on minimum and maximum values. In this context, the work of Antoniou et al. (2019) employed automated planning to optimize the length of traffic signal phases by leveraging the existing physical infrastructure, eliminating the need for additional hardware; in simulation, the approach was demonstrated to be capable of reducing congestion and emissions by up to 20%. Automated planning constitutes a foundational discipline of artificial intelligence, and it is concerned with the task of generating plans (sequences of actions) based on descriptions of a system to be controlled and its dynamics, initial states, and specified goals to be achieved (Ghallab et al., 2004; 2016). A knowledge model is devised to provide the essential information required to resolve the corresponding task. This knowledge model encompasses predictive details about actions and specifically outlines how these actions alter the environment's state upon application (McCluskey et al., 2017). Various formalisms exist to articulate this knowledge, with the widely recognized Planning Domain Description Language (PDDL) being prominent.

The work of Antoniou et al. has then been extended in Percassi et al. (2023), El Kouaiti et al. (2024), where the proposed AI approaches have been extensively validated with historical data from urban regions in Manchester and Huddersfield, from the northern part of the United Kingdom. These most recent works take into account the constraints of the existing infrastructure, and are hence suitable for deployment. Real-world deployment of the proposed techniques has been done in a number of trials in Huddersfield and in Hull, with remarkably positive impact on traffic congestion and journey times. In the former case, the AI-based approach has been used to ease the traffic conditions after a major music event in Huddersfield, while in the latter case, it has been used for a week to control traffic during morning and evening peak hours in Hull. It is worth noting that the deployments highlighted one of the main challenges of applying AI techniques to the traffic signal optimization problem, which is the access to real-world data. Many traffic authorities cannot afford to buy and install dedicated sensing infrastructure, hence the need to exploit available sensors and systems in operation. However, data generated from different sources can have different granularity, might require different interpretation, and might not be available at the same time, or even within a reasonable time horizon to be used (Bhatnagar et al., 2022b). The availability of data is a crucial aspect for most AI applications, and in the considered line of work, the specific issues forced a complete redesign of the framework and of the mode of operations.

One notable benefit of using automated planning is that it is possible to leverage on the knowledge models needed by the AI system to create accurate and easily replicable simulators (Bhatnagar et al., 2022a; 2023), as well as to support what-if analysis and in-detail investigations of the generated strategies (Vallati and Chrpa, 2023; Percassi and Vallati, 2024). For the first time, by means of model-based AI, many traffic authorities have access to reliable and inexpensive simulators to be used in their region, and can explore the expected impact of the strategies they are going to implement in a principled way.

Considering a different topic within the field of traffic control and mobility, artificial intelligence has also been widely used for addressing traffic congestion through urban traffic routing and traffic signal control. Its ability in processing and analyzing vast amounts of data enables continuous monitoring of traffic conditions, predicting congestion patterns, and dynamically adjusting routing strategies to respond to traffic fluctuations. This adaptability allows for smoother traffic flows in dynamic routing scenarios (Guo et al., 2023b). For example, in the study by Dikshit et al. (2023), AI-driven data analysis and algorithms were applied to optimize dynamic routing, reducing commuter travel times and promoting a more sustainable urban traffic environment. Additionally, automated planning, a research field of AI, has shown its advantages in traffic routing and distribution, especially in centralized systems. The work of Švadlenka et al. (2023) highlighted the scalability of automated planning in planning-based traffic routing, incorporated A* search algorithm for generating routes and DBSCAN for route clusters. The work of Cardellini et al. (2024) showed the impact of centralized distribution of vehicles in cities with different network structures, highlighting the importance of network topology for the successful application of these kinds of approaches.

AI applications in public transport (PT) services have become a promising way of improving system efficiency and quality of service (QoS) (Guo, 2024). Compared to traditional PT operations (including route planning, timetable design, and vehicle and crew scheduling), AI-driven service optimization techniques have increasingly enhanced the performance of PT services (Buba and Lee, 2018). For example, heuristic algorithms like genetic algorithms (GA), simulated annealing (SA), particle swarm optimization (PSO), and hybrid approaches have been applied to address multi-objective optimization problems for bus routing and scheduling. Examples can be seen in Yao et al. (2014) and Buba and Lee (2018). These techniques help balance the trade-off between operators and passengers by maximizing operating costs, passenger satisfaction, or minimizing vehicle utilization, thus eventually improving accessibility, mobility, and reducing waiting times, travel times, and transfers (Feng et al., 2019). AI solutions have also played a significant role in addressing transit service design challenges for electric-based buses. Advanced algorithms are employed to jointly optimize routing, scheduling, and charging operations considering different charging modes and behaviors, helping to minimize charging costs while maximizing service coverage, such as the work of Zhou et al. (2022). Emerging Connected and Autonomous Vehicles (CAVs) have further extended the possibilities for AI applications in PT. CAV technology has the capability to potentially enhance PT systems by offering real-time data collection, dynamic route adjustments, and improved fleet management (Azad et al., 2019). Autonomous Public Transport (APT) systems, for instance, use AI approaches for automated scheduling and advanced vehicle assignment, which improves vehicle utilization and energy efficiency, refer to Cao and Ceder (2019). Meanwhile, demand-responsive transit (DRT) systems, empowered by AI and CAVs, can adaptively cater to passenger demand fluctuations in real-time, offering tailored services such as door-to-door on-demand buses without transfers. Such successful implementation can be seen in Wu et al. (2021) and Gong et al. (2021). Integrating CAVs into DRT networks further enables innovations like modular autonomous vehicles, where autonomous vehicles can be assembled or disassembled into modular units based on dynamic passenger travel requirements, allowing for more adaptable and flexible capacity management. AI thus offers promising solutions to optimize PT systems, balancing operational efficiency with enhanced passenger experiences (Nikitas et al., 2020; Wang et al., 2022).

9.3 Textile Sorting

The textile industry is a significant contributor to environmental pollution and degradation, responsible for 10% of global carbon emissions. Beyond its carbon footprint, the industry currently produces over 92 million tonnes of textile waste annually, with this figure expected to reach 134 million tonnes by 2030 (Chen et al., 2021). This is largely driven by the fast fashion trend. Fast fashion, characterized by cheap, disposable clothing, encourages consumers to frequently purchase new items, which often quickly end up as waste (Niinimäki et al., 2020).

A major component of this waste comes from synthetic polymers, such as polyester, nylon, and acrylic, which are largely used in textile production. These materials are not biodegradable and, as such, are a major source of plastic pollution. For instance, in 2017, the textile industry accounted for 62 million tonnes of plastic waste, contributing substantially to the global total of 438 million tonnes (Pensupa et al., 2018).

Despite the rising awareness around plastic waste, much of the attention has historically been focused on plastic packaging, such as bottles, bags, and wrappers, with its visibility in landfills and oceans, and environmental impact well documented (MacArthur, 2017). Only recently has textile waste garnered more attention, thanks to the rapid growth of the textile industry and, consequently, the rapid growth in its environmental impact. However, dealing with textile waste presents complex challenges when it comes to recycling.

Recycling textiles is complicated due to the nature of the materials involved. Many garments today are made from blends of both synthetic and natural fibers, for example, cotton mixed with polyester. It is necessary to separate these fibers prior to recycling because different fiber types require different recycling processes. Chemical approaches are being designed to remove natural fibers to allow separation and re-use of the synthetic ones (Hu et al., 2018), but they are only effective if textiles are correctly sorted by fiber composition, as the chemical approach to be used and optimized depends on the type and percentage of fibers present.

At present, a vast amount of textile waste is discarded rather than being reused or recycled due to the absence of reliable sorting processes. Much of the sorting in textile recycling facilities is currently done manually by human operators. This process is not only labor-intensive and time-consuming, but also prone to errors. Workers require months of training to properly identify and classify different fiber types. The cost of labor and the inherent inaccuracies of manual sorting have created significant barriers to improving textile recycling efforts on a larger scale.

As part of the EPSRC UK "Textile Waste Refinery for the production of recycled plastic, cellulose and dye" project, we are investigating the use of machine learning for automated sorting of postconsumer textiles based on fiber type. Our initial focus is on distinguishing cotton and polyester fabrics using spectrophotometric analysis. This rapid, nondestructive method measures light absorption and provides valuable data for AI model training, paving the way for efficient, automated textile waste recycling.

In a preliminary empirical evaluation (Robinson et al., 2024) we trained four well-known machine learning approaches on datasets composed of features extracted by spectrophotometric analysis on pure cotton or polyester samples in a limited range of colors, and achieved promising results. In particular, we observed that an accuracy of approximately 90% can be achieved on test samples that are of similar color to those used in the training, with preliminary results suggesting that the techniques can generalize when a larger set of samples is considered. We are now in the process of investigating performance on larger datasets, and extending the capabilities of the proposed approaches in the presence of blended fibers, where recycling issues are exacerbated.

This work is a step forward toward the recycling of textile waste, and shows how AI and other disciplines can come together to provide encompassing solutions to challenging problems.

9.4 Power Distribution Networks

Power distribution networks have been developed to transport electricity from the transmission network to end users in urban and rural areas. In France, the total length of the distribution network is 1 377 269 km compared to 105 942 km for the transmission network. Given the scale of the network, any investment has to be fully justified by a thorough technical and economic assessment (Gönen et al., 2024).

The importance of these networks has become even more critical as new energy challenges emerge, such as the growing demand for electricity, the integration of intermittent renewable energy sources, and the desire to reduce greenhouse gas emissions. In this context, the integration of smart grids is changing the way electricity distribution networks operate. They include sensors and real-time data transmission devices installed on the various network grids, allowing the state of the network to be analyzed and monitored. However, they face many changes, which have started to emerge in recent years at different speeds depending on the country (Colak et al., 2020).

The distribution network was not originally designed to accommodate generation units. Rather, it was designed as a centralized and unidirectional system, with the objective of transporting electricity from the distribution network to medium and low-voltage consumers. In the current context, however, it must integrate energy production that is distributed throughout the country. This production is, in some cases, intermittent, following the gradual integration of renewable energy sources, including wind and photovoltaic energy.

In order to anticipate these significant changes, distribution grid operators (hereinafter, DSOs) aim to define the optimal long-term grid architecture (in a horizon of a few decades) called the target grid. This is done by optimizing a set of technical and economic performance indicators while respecting a set of constraints that have been predefined. Such modifications may be *minor*, by adding and/or removing lines and transformers, or *intermediate*, by creating new parts connected to the existing grid (expansion planning) or *major*, by changing the complete architecture (green-field planning) (Khator and Leung, 1997); Temraz and Quintana, 1993). Once the target grid has been determined, the set of intermediate grids allowing the transition from the initial grid to the target grid must be defined. The aforementioned set can be complex, especially if the network is large. The mission of DSOs is to plan the work in the short term based on technical and economic calculations. If the previous calculations are done with the help of software, the planning work itself, i.e. imagining the network topology and the evolution strategies, is done by hand relying on the expertise of the DSOs.

The decision support tool for DSOs presented in Castellanos et al. (2023) demonstrates an efficient solution to the aforementioned problem of finding intermediate networks through the integration of AI techniques based on Automated Planning and electrical computations. Indeed, AI planning is beneficial to address problems subject to continuous change (e.g. integration of renewable energies, management of the load caused by electric vehicles, etc.). A large number of high-performance planners (implementing different approaches) are already available, and planning algorithms are constantly evolving. The experimental analysis of that work, on several networks at different scales, demonstrated the applicability of the approach and the reduction in reliance on expert knowledge. Furthermore, the solution found by a smart grid expert required one hour, whereas the AI solution required only 0.07 seconds.

Moving on from the planning of the grid architecture (long-term planning) to the optimization of its operation (on a daily scale), it is necessary to address the issue of the feed-in from renewable energy sources (RES) being difficult to influence due to its weather-dependent generation. In order to guarantee the voltage quality, the safe and reliable operation, and to avoid costly expansion of the electric grids, volatile RES, distributed battery storage units, and controllable loads must be intelligently and controllably connected to each other.

The bilateral Franco-German project AI4DG[1] aims to research and develop a distributed AI on the edge platform for an autonomous and secure battery storage

---

1 https://anr.fr/Project-ANR- 21-FAI2-0008

control system in low-voltage grids with a high share of renewable energy sources using smart meter data. Due to the distributed AI on the edge approach, the system is more fail-safe and able to preprocess sensible smart meter data to ensure data protection regulations than a centralized approach. In this ongoing project, AI is used as a prediction module in the day-ahead house power optimizer. Machine learning models are built based on the WWN dataset (field test). It consists of data collected on a low-voltage distribution network operated by one of the project partners in Germany—Westfalen Weser Netz GmbH (WWN)—for four residential houses connected to the distribution network and equipped with smart meters.

This work is ongoing, and the next steps will include implementing the prediction and control strategies in an edge platform.

9.5 Discussion and Conclusion

In this chapter, we presented the AI-based solutions designed to deal with urban mobility and traffic control, post-user textile sorting, and power distribution networks. While the focus has been on urban mobility and traffic control, that is the most mature area of application, it should be noted that in all cases the corresponding projects are still undergoing, hence further progress in the near future is expected. In our analysis, we focused on the main challenges faced in each domain, and on the impact of the AI approaches, with particular attention given to the environmental dimensions.

The results already achieved by the AI techniques designed for the case studies can be attributed to several key factors.

- First, the significant involvement of domain experts in the research process. In many domains, experts possess a vast amount of domain knowledge that needs to be elicited and encoded to allow AI approaches to operate efficiently and effectively. It is worth noting that, at a time when Large Language Models are seen as the de facto AI approach to solve any problem (Hadi et al., 2024), our case studies shed some light on the fact that in challenging application domains there is still the need for (i) human experts' involvement and (ii) selection of the most suitable AI approaches that support validation and verification of the acquired knowledge and of the overall operations of the AI system (Vallati and Chrpa, 2023).
- Second, the approach based on rapid prototyping and fast feedback loops facilitated quick iterations and continuous improvement, so that the AI solutions were refined and optimized over time. This iterative process allowed for the early identification and mitigation of potential challenges, and ensured that AI approaches were solving the right problem in the right way.

- Third, the importance of taking into account potential threats and vulnerabilities that an autonomous AI system may pose. Focusing on the transport sector, significant work is now being done to raise awareness around the potential for unprecedented attacks to vehicles and to infrastructure (Nikitas et al., 2022; Parkinson et al., 2024; Guo et al., 2023a): this attention needs to be transferred to other domains sharing similar challenges, such as smart grids.

We should now turn our attention to the main challenges faced in the case studies and those that are likely to be faced in a range of application domains. First, it is crucial to consider the deployment infrastructure. While existing systems and infrastructure may have been designed for human operators or traditional software, AI systems often function in fundamentally different ways. This divergence can lead to hidden constraints or issues that are not immediately apparent. As an exhaustive assessment of the infrastructure is usually infeasible or impossible, AI systems should be designed with flexibility and robustness in mind. Additionally, there is a significant need for transparency and explainability of AI approaches exploited in critical applications. Explainability has to be defined on the basis of the users and domain, but human experts should be in the position to query the AI system and gain an understanding of the way in which a given outcome has been generated. Finally, the availability of high-quality data is a well-known prerequisite for a wide range of AI approaches, particularly those based on machine learning techniques.

Finally, it is crucial to take into proper account the environmental impact of AI solutions (Ligozat et al., 2021; Delort et al., 2023): the design, development, and optimization of AI approaches come at a potentially high environmental cost, due to the need to run massive and computationally expensive processes. This cost is exacerbated in deep learning cases where huge amount of data is needed to train the initial model and to support subsequent retraining to adapt models to changed circumstances. Also, in the case of model-based approaches, the cost of automated reasoning on the fly can be significant, and must be accounted for to develop truly sustainable solutions.

Bibliography

Grigoris Antoniou, Sotiris Batsakis, John Davies, Alistair Duke, Thomas L McCluskey, Evtim Peytchev, Ilias Tachmazidis, and Mauro Vallati. Enabling the use of a planning agent for urban traffic management via enriched and integrated urban data. *Transportation Research Part C: Emerging Technologies*, 98:284–297, 2019.

Mojdeh Azad, Nima Hoseinzadeh, Candace Brakewood, Christopher R Cherry, and Lee D Han. Fully autonomous buses: A literature review and future research directions. *Journal of Advanced Transportation*, 2019:1–16, 2019.

Saumya Bhatnagar, Rongge Guo, Keith McCabe, Thomas L McCluskey, Enrico Scala, and Mauro Vallati. Leveraging artificial intelligence for simulating traffic signal strategies. In *2022 IEEE 25th International Conference on Intelligent Transportation Systems (ITSC)*, pp. 607–612. IEEE, 2022a.

Saumya Bhatnagar, Sumit Mund, Enrico Scala, Keith McCabe, Thomas Leo McCluskey, and Mauro Vallati. On-the-fly knowledge acquisition for automated planning applications: Challenges and lessons learnt. In *Proceedings of the 14th International Conference on Agents and Artificial Intelligence, ICAART, Scitepress*, pp. 387–397, 2022b.

Saumya Bhatnagar, Rongge Guo, Keith McCabe, Thomas L McCluskey, Francesco Percassi, and Mauro Vallati. Automated planning for generating and simulating traffic signal strategies. In *Proceedings of the Thirty-Second International Joint Conference on Artificial Intelligence, IJCAI*, pp. 7119–7122, 2023.

Ahmed Tarajo Buba and Lai Soon Lee. A differential evolution for simultaneous transit network design and frequency setting problem. *Expert Systems with Applications*, 106:277–289, 2018.

Zhichao Cao and Avishai Avi Ceder. Autonomous shuttle bus service timetabling and vehicle scheduling using skip-stop tactic. *Transportation Research Part C: Emerging Technologies*, 102:370–395, 2019.

Matteo Cardellini, Carmine Dodaro, Marco Maratea, and Mauro Vallati. Optimising dynamic traffic distribution for urban networks with answer set programming. *Theory and Practice of Logic Programming*, 24(4):1–19, 2024. https://doi.org/10.1017/S1471068424000309.

Sandra Castellanos, Marie-Cécile Alvarez-Herault, and Philippe Lalanda. Decision support tool for the development of power distribution networks based on AI planning. In *27th International Conference on Electricity Distribution (CIRED 2023)*, volume 2023, pp. 1425–1429. IET, 2023.

Xuandong Chen, Hifza A Memon, Yuanhao Wang, Ifra Marriam, and Mike Tebyetekerwa. Circular economy and sustainability of the clothing and textile industry. *Materials Circular Economy*, 3:1–9, 2021.

Ilhami Colak, Ramazan Bayindir, and Seref Sagiroglu. The effects of the smart grid system on the national grids. In *2020 8th International Conference on Smart Grid (icSmartGrid)*, pp. 122–126, IEEE, 2020.

Etienne Delort, Laura Riou, and Anukriti Srivastava. *Environmental Impact of Artificial Intelligence*. PhD thesis, INRIA, pp. 1–33. CEA Leti, 2023.

Srishti Dikshit, Areeba Atiq, Mohammad Shahid, Vinay Dwivedi, and Aarushi Thusu. The use of artificial intelligence to optimize the routing of vehicles and reduce traffic congestion in urban areas. *EAI Endorsed Transactions on Energy Web*, 10, 2023.

Anas El Kouaiti, Francesco Percassi, Alessandro Saetti, Thomas Leo McCluskey, and Mauro Vallati. PDDL+ models for deployable yet effective traffic signal optimisation. In *Proceedings of the International Conference on Automated Planning and Scheduling*, volume 34, pp. 168–177, AAAI Press, 2024.

Xuesong Feng, Xiaojing Zhu, Xuepeng Qian, Yuanpeng Jie, Fei Ma, and Xuejun Niu. A new transit network design study in consideration of transfer time composition. *Transportation Research Part D: Transport and Environment*, 66:85–94, 2019.

Malik Ghallab, Dana Nau, and Paolo Traverso. *Automated Planning: Theory and Practice*. Elsevier, 2004.

Malik Ghallab, Dana S Nau, and Paolo Traverso. *Automated Planning and Acting*. Cambridge University Press, 2016.

Turan Gönen, Chee-Wooi Ten, and Ali Mehrizi-Sani. *Electric Power Distribution Engineering*. CRC Press, 2024.

Manlin Gong, Yucong Hu, Zhiwei Chen, and Xiaopeng Li. Transfer-based customized modular bus system design with passenger-route assignment optimization. *Transportation Research Part E: Logistics and Transportation Review*, 153:102422, 2021.

Rongge Guo. AI approaches on urban public transport routing. In *Deception in Autonomous Transport Systems: Threats, Impacts and Mitigation Policies*, pp. 111–130. Springer, 2024.

Rongge Guo, Alexandros Nikitas, Simon Parkinson, and Mauro Vallati. One bus is all it takes: The impact of cyber attacks in customised bus services. In *The IEEE International Conference on Intelligent Transportation Systems, ITSC*, pp. 5784–5789. IEEE, 2023a.

Rongge Guo, Mauro Vallati, Yutong Wang, Hui Zhang, Yuanyuan Chen, and Fei-Yue Wang. Sustainability opportunities and ethical challenges of ai-enabled connected autonomous vehicles routing in urban areas. In *IEEE Transactions on Intelligent Vehicles*, pp. 55–58. IEEE, 2023b.

Muhammad Usman Hadi, Qasem Al Tashi, Abbas Shah, Rizwan Qureshi, Amgad Muneer, Muhammad Irfan, Anas Zafar, Muhammad Bilal Shaikh, Naveed Akhtar, Jia Wu, et al. Large language models: A comprehensive survey of its applications, challenges, limitations, and future prospects. *Authorea Preprints*. TechRxiv, 2024.

Yunzi Hu, Chenyu Du, Nattha Pensupa, and Carol Sze Ki Lin. Optimisation of fungal cellulase production from textile waste using experimental design. *Process Safety and Environmental Protection*, 118:133–142, 2018.

Suresh K Khator and Lawrence C Leung. Power distribution planning: A review of models and issues. *IEEE Transactions on Power Systems*, 12(3):1151–1159, 1997.

Jonathan I Levy, Jonathan J Buonocore, and Katherine Von Stackelberg. Evaluation of the public health impacts of traffic congestion: A health risk assessment. *Environmental Health*, 9:1–12, 2010.

Anne-Laure Ligozat, Julien Lefèvre, Aurélie Bugeau, and Jacques Combaz. Unraveling the hidden environmental impacts of AI solutions for environment. *arXiv preprint arXiv:2110.11822*, 2021.

Ellen MacArthur. Beyond plastic waste. *Science*, 358(6365):843, 2017.

Thomas Leo McCluskey, Tiago Stegun Vaquero, and Mauro Vallati. Engineering knowledge for automated planning: Towards a notion of quality. In *Proceedings of the Knowledge Capture Conference, K-CAP*, 14:1–8, 2017.

Kirsi Niinimäki, Greg Peters, Helena Dahlbo, Patsy Perry, Timo Rissanen, and Alison Gwilt. The environmental price of fast fashion. *Nature Reviews Earth & Environment*, 1(4):189–200, 2020.

Alexandros Nikitas, Kalliopi Michalakopoulou, Eric Tchouamou Njoya, and Dimitris Karampatzakis. Artificial intelligence, transport and the smart city: Definitions and dimensions of a new mobility era. *Sustainability*, 12(7):2789, 2020.

Alexandros Nikitas, Simon Parkinson, and Mauro Vallati. The deceitful connected and autonomous vehicle: Defining the concept, contextualising its dimensions and proposing mitigation policies. *Transport Policy*, 122:1–10, 2022.

Simon Parkinson, Alexandros Nikitas, and Mauro Vallati. *Deception in Autonomous Transport Systems: Threats Impacts and Mitigation Policies*. Springer, 2024.

Nattha Pensupa, Shao-Yuan Leu, Yunzi Hu, Chenyu Du, Hao Liu, Houde Jing, Huaimin Wang, and Carol Sze Ki Lin. Recent trends in sustainable textile waste recycling methods: Current situation and future prospects. *Topics in Current Chemistry*, 375(5):76, 2018.

Francesco Percassi and Mauro Vallati. Leveraging AI planning in a what-if analysis framework for assessing traffic signal strategies. In *27th IEEE International Conference on Intelligent Transportation Systems*, pp. 1330–1335. IEEE, 2024.

Francesco Percassi, Saumya Bhatnagar, Rongge Guo, Keith McCabe, Thomas L Mccluskey, and Mauro Vallati. An efficient heuristic for ai-based urban traffic control. In *2023 8th International Conference on Models and Technologies for Intelligent Transportation Systems (MT-ITS)*, pp. 1–6. IEEE, 2023.

Robinson, Megan Ghosh, Saikat Goswami, and Parikshit Vallati. Mauro A novel methodology for ai-based sorting of post-consumer textile using spectrophotometer. In *1st International Workshop on Artificial Intelligence for Climate Change*, pp. 1–8, 2024.

Matyáš Švadlenka, Lukáš Chrpa, and Mauro Vallati. Improving the scalability of automated planning-based vehicle routing via smart routes identification. In *2023 8th International Conference on Models and Technologies for Intelligent Transportation Systems (MT-ITS)*, pp. 1–6. IEEE, 2023.

Hesham K Temraz and Victor H Quintana. Distribution system expansion planning models: An overview. *Electric Power Systems Research*, 26(1):61–70, 1993.

Mauro Vallati and Lukáš Chrpa. In defence of good old-fashioned artificial intelligence approaches in intelligent transportation systems. In *2023 IEEE 26th International Conference on Intelligent Transportation Systems (ITSC)*, pp. 4913–4918. IEEE, 2023.

Zhiwen Wang, Kangkang Yang, Long Li, Yanrong Lu, and Yufei Tao. Traffic signal priority control based on shared experience multi-agent deep reinforcement learning. *IET Intelligent Transport Systems*, 17(7):1363–1379, 2022.

Jiaming Wu, Balázs Kulcsár, Selpi, and Xiaobo Qu. A modular, adaptive, and autonomous transit system (MAATS): An in-motion transfer strategy and performance evaluation in urban grid transit networks. *Transportation Research Part A: Policy and Practice*, 151:81–98, 2021.

Baozhen Yao, Ping Hu, Xiaohong Lu, Junjie Gao, and Mingheng Zhang. Transit network design based on travel time reliability. *Transportation Research Part C: Emerging Technologies*, 43:233–248, 2014.

Yu Zhou, Qiang Meng, and Ghim Ping Ong. Electric bus charging scheduling for a single public transport route considering nonlinear charging profile and battery degradation effect. *Transportation Research Part B: Methodological*, 159:49–75, 2022.

10

Blockchain-enabled Digital Twins for Advancing Sustainable Reverse Logistics in Renewable Energy Systems

*Deepak Gautam**

School of Management and Entrepreneurship, Shiv Nadar Institution of Eminence, Noida, Uttar Pradesh, India

*Corresponding author: deepakgautam19932@gmail.com

10.1 Introduction

In the past few years, there has been an enormous push toward green energy. People are very worried about climate change, running out of fossil fuels, and the need for better, longer-lasting energy options. To meet their lofty decarbonization goals, governments, businesses, and people are using more and more renewable technologies, such as wind farms, solar panels, and electric cars (Di Silvestre et al., 2018). Renewably, these systems become cornerstones of the global energy mix as they contribute meaningfully to cuts in greenhouse gas emissions and support a greener, more ecologically friendly energy-use revolution.

Without doubt, rapid and widespread development of renewable energy systems will throw up new kinds of problems. Of these is perhaps the largely overlooked, although very crucial, proper end-of-life (EoL) cycle system management, comprising decommissioning, recycling, reuse, or eventual disposal of all components with EoL (Bouzarour-Amokrane et al., 2015). Without a sound infrastructure set up to govern these processes, such fixes intended to limit environmental degradation may themselves become sources of pollution, waste, and inefficiency (Omer, 2008).

Reverse logistics is used in this process. It is the process of getting items back from customers to the person who made them or to the next step in the value chain so that they can be recycled, fixed up, or thrown away properly (Wilson and

Renewable Energy Transition with Artificial Intelligence: Challenge-driven Solutions, First Edition. Nina Dethlefs and Joyjit Chatterjee. Published 2026 by John Wiley & Sons, Inc.

Goffnett, 2022). In green energy, reverse logistics is the process of getting things like wind turbines, solar panels, and EV batteries back so they can be reused, recycled, or given to someone else. The importance of reverse logistics especially becomes very apparent in the context of the circular economy, a system that has been engineered to keep products, components, and materials in use for as long as possible, thereby keeping waste and virgin resource demand to a minimum (Butt et al., 2024).

Reverse logistics is a very critical input into sustainability in renewable energy; however, it is complex. Challenges facing the renewable energy industry include:

- The data relating to the health, usage, or ownership of assets is fragmented, implying that there are bad decisions (Psara et al., 2022).
- The state and lifetime of green energy assets are usually not visible in the value chain, so it becomes difficult to track the best ways to resell or recycle them (Yu et al., 2023).
- If there isn't a tracking system to see the real-time status of assets, then recycling programs mightn't happen on a regular plan, and so may incur extra costs for storage or destruction.

Addressing these issues is necessary to make renewable energy systems not just carbon neutral during their working lives but also sustainable in a life-cycle sense. This book chapter introduces a new response to these challenges: blockchain-enabled digital twins (BEDT). Taking advantage of blockchain technology as well as digital twin (DT) models, this framework aims to provide optimal reverse logistics processes in the context of renewable energy assets and promote greater transparency, efficiency, and sustainability in all life-cycle aspects.

10.1.1 Reverse Logistics Importance in Renewable Energy

Renewable energy systems have long been touted for their clean, low-emission electricity production. Still, most sustainability discussions tend to focus on the operational phases of such systems-how much electricity they produce, their carbon offsets, and their reduction in dependence on fossil fuels. While these metrics are undoubtedly key, sustainability in renewable energy systems also has to consider how their parts are postoperationalized.

For example, the design lifetime for a solar PV panel is intended to last from 25 to 30 years (Walker and Casey, 2024). The latter half of this century will approach nearly 78 million metric tons of waste for such panels' proper EoL management, that's likely to be an important component of environmental pollution once more, again antithetical to what solar energy would otherwise purport to represent (Ali et al., 2024). The same applies to the wind turbine blade, mostly made up of composites, which has remained nonrecyclable and ended up primarily in landfills after retirement. Estimates from the International Energy Agency show that around 43 million tons of wind turbine blades are projected to retire by 2050 across the world (Gao, 2023). Issues relating to the EoL cycle for EV battery life are equally as pertinent as the issues discussed above for equivalent reasons associated with electric mobility. This projected 1.2 million tons of EV batteries will need to be recycled or repurposed by 2030 (Zhao et al., 2021).

As shown in the above examples, it reflects the magnitude of the task. Unless properly controlled, renewable energy systems will develop substantial waste streams. And the solution is by establishing reverse logistics processes that support reuse, refurbishment, and recycling of parts so the lifespan of the parts will increase and will be supportive for a circular economy. Reverse logistics in renewable energy is not only a need to reduce environmental impacts but also an opportunity for business because materials like lithium, cobalt, and rare earth elements, which are part of the EV battery and other renewable energy systems, can be recycled and reused instead of relying on mining for new materials (Gautam and Bolia, 2024).

10.1.2 Blockchain-enabled Digital Twins

With a focus on overcoming the obstacles that reverse logistics of renewable energy impose on the system, the resolution lies in integrating asset life-cycle data and transparent treatment. This book presents blockchain-enabled digital twins (BEDT) as a conceptual framework aligned with those objectives.

10.1.2.1 Digital Twins

A digital twin (DT) is a real-time representation of a physical asset, system, or process (Dietz and Pernul, 2020). In the context of renewable energy, DTs could mimic live situations as in wind turbines, solar panels, and EV batteries, thereby giving live insights into their operational conditions, health, and usage history (Song et al., 2022). The continuous monitoring and analysis supported by DTs enable stakeholders to optimize when to maintain, recycle, or repurpose. The DT mirrors the physical asset in real time, providing dynamic information to support critical reverse logistics decision-making (Kamble et al., 2022).

A DT of a wind turbine blade, for example, would enable the monitoring of real-time wear and tear in such a blade, which could make predictions about whether to replace or refurbish it (Chetan et al., 2021). For solar panels, it can also be used for monitoring degradation over time so as to schedule its recycling or disposal precisely (Fan and Li, 2023). General Electric has already applied DT technology in its wind turbines to monitor their performance and align optimal maintenance schedules for better operational efficiency (Ismail et al., 2024).

10.1.2.2 Blockchain

Blockchain technology in reverse logistics is able to enhance the possibility of DTs for providing real-time insights. It is a decentralized ledger that cannot be tampered with and records all the transactions and data exchanges connected to an asset. Because of this, the DT will make sure that the data is safe, can be tracked, and is clear throughout the whole supply chain (Bakhshi et al., 2024).

In the case of reverse logistics in green energy, blockchain makes sure that everyone involved—whether they are the maker, a support provider, a collector, or the government—always has access to correct information about the life cycle of an asset. Transparency is one of the strong attributes of blockchain that would catalyze environmental regulation compliance and stakeholder trust (Acosta Llano et al., 2024). Consider the owner's history and usage conditions for a battery from an electric vehicle (EV); through blockchain, all that information can be recorded to make it easier to decide whether it is fit for further use or recycling.

This way, including data produced by DTs of real time, as a result of transparent, safe data management brought about through blockchain, the entire framework of BEDT supplies the user with one complete and integrated system capable of optimization processes for reverse logistics of systems in renewable energy.

10.1.3 Technical Architecture of BEDT

The BEDT framework combines blockchain, digital twins, AI, and IoT into a streamlined system to enhance reverse logistics for renewable energy assets like wind turbines, solar panels, and EV batteries. It features key components: physical assets monitored by IoT sensors, digital twins that mirror these assets using real-time data, AI algorithms that predict maintenance or EoL, a blockchain ledger for secure and transparent data storage, and stakeholder interfaces for decision-making. The process starts with IoT sensors collecting data on asset conditions, which updates the digital twin. AI then analyzes this data to forecast asset health, and critical insights are logged on the blockchain. Stakeholders use this information to decide on maintenance, recycling, or repurposing, ensuring a transparent, efficient, and sustainable approach to managing renewable energy life cycles.

The BEDT predictive asset management framework architecture is depicted in Figure 10.1. The framework incorporates several components, such as IoT sensors for real-time data capture, cloud storage for off-chain data management, and a blockchain network for on-chain data validation and transparency. IoT sensors

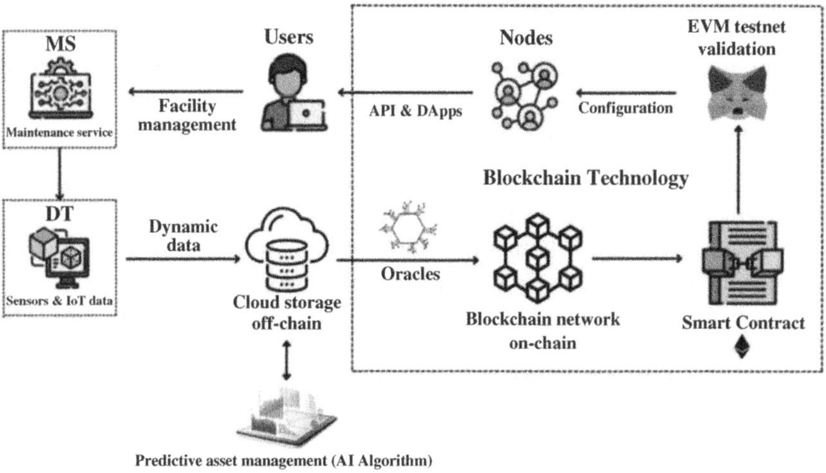

Figure 10.1 Architecture of BEDT framework (Tavakoli et al., 2024/Emerald Publishing Limited/CC BY 4.0).

record dynamic data from physical equipment, which is processed by digital twin models to generate real-time simulations. Blockchain technology, with smart contracts driving it, provides secure and tamper-evident storage of data, and oracles enable data transfer between off-chain and on-chain. Predictive asset management is enabled by AI algorithms that cross-reference the data to optimize maintenance schedules, minimize downtime, and maximize resource utilization. The collaborative strategy enables green reverse logistics by way of providing smart decisions in real time to the stakeholders regarding asset condition and life-cycle management.

This framework is in line with the main agendas of this chapter, which are centered on tapping into BEDT to optimize renewable energy systems' sustainable reverse logistics. With IoT sensors, blockchain, and artificial intelligence (AI)-based predictive analytics integrated within it, the framework provides a comprehensive solution to real-time tracking, open-data management, and best-informed decision-making from the cradle to the grave of renewable energy assets. The figure illustrates the ability of BEDT to address some of the greatest challenges in reverse logistics, such as disintegrated data, traceability concerns, and waste recycling and refurbishment operation inefficiencies. The figure also reflects the possibility of this model toward enhancing compliance with environmental regulations as well as refining circular economy thinking, paving the way toward an enhanced, sustainable, and resource-efficient renewable energy supply chain.

10.1.4 Chapter Objectives

The core objective of this chapter will be to present an articulated conceptual framework for BEDT in the context of reverse logistics for renewable energy systems. In particular, this chapter will:

- Explain what DTs and blockchain technology are and how they can be used together to make reverse operations better.
- Show how the BEDT framework can be used in key areas of green energy, such as battery systems for EVs, wind farms, and solar panels.
- Use case studies from the green energy business to show how this framework can be used in the real world.
- Explain how BEDT contributes to the shift toward a more sustainable, open, and efficient economy.

10.2 Digital Twins and Blockchain in Reverse Logistics

In the line of integration, DT technology is merged with blockchain to bring about a more innovative approach toward reverse logistics, especially renewable energy systems. It is a unique approach toward life-cycle management, especially concerning the whole concept of digital twins, which create real-time virtual models of physical assets, and blockchains' secure, decentralized ledger. These technologies, when taken together, provide accurate, transparent, and traceable data in support of more sustainable and efficient handling of assets such as wind turbines, solar panels, and EV batteries from production to the EoL stage (Hakimi et al., 2023).

DTs replicate the conditions that exist in the real world; hence, it senses the performance, health, and conditions of assets and the environment to simulate proactive decision-making. The following are some of the notable applications that show versatility and industry-wide impacts of DTs in renewable energy:

- Integration in Monitoring Wind Turbines: General Electric and Siemens Gamesa produced some of the first machines that had DTs fitted into them (Eleftheriou and Anagnostopoulos, 2022). The DTs enable monitoring of blade wear and tear, motor performance, and operational efficiency, thus enabling maintenance teams to schedule proper repair or replacement at an appropriate time. Such an approach saves their wind turbines from untimely breakdown while maximizing their operational life cycle, curtailing a significant level of operational downtimes and waste (Chetan et al., 2021).
- The leading manufacturer of solar panels, First Solar, employs DTs to monitor panel efficiency, degradation, and exposure to environmental factors through digital models (Chowdhury et al., 2020). This enables the maintenance teams to identify when panels require cleaning and repairs, or even a replacement.

It is possible to optimize production in real time for every panel while minimizing resource use and waste with the help of real-time assessment of every panel by First Solar.

- In managing the entire life cycle of electric vehicle batteries, the companies use DTs for management starting from manufacturing till the EoL. Using such technology, the performance, charging cycles, and temperature fluctuations of the battery are monitored to allow companies to forecast when exactly it should be recycled or repurposed (Rajesh et al., 2024).

All these applications illustrate how DTs empower renewable energy companies to track their assets in real time to support life-cycle decisions based on circular economy principles. It will finally bring all the imperative benefits through improvements in operational efficiency and better resource allocation toward more sustainable practices across the industry.

10.2.1 Blockchain and Transparency and Traceability

Blockchain technology plays a fundamental role in clarifying, securing, and auditing reverse logistics processes (Acosta Llano et al., 2024; Srivastava et al., 2021). As a tamper-proof, decentralized ledger, blockchain captures every data point along an asset's life cycle and provides its diverse stakeholders—manufacturers, recyclers, regulators, and end-users—with accurate information in real time (Brandín and Abrishami, 2021). Core advantages of blockchain in reverse logistics are transparency and traceability, which would provide trust and compliance with the supply chain.

Blockchain has been used in two real applications that bring transparency and traceability to renewable energy. Some of them include:

- Volkswagen and IBM's Blockchain Pilot for EV Batteries: Volkswagen collaborated with IBM as part of a partnership, initiating a blockchain pilot to track the life cycle of EVs from cradle to grave. The question of transparency in the supply chain has become very critical through this application, since it establishes a safe and decentralized ledger that stores any condition change, transaction, or ownership transfer, up to recycling history related to the battery. With blockchain immutability, participants—the manufacturers, recyclers, and even the regulators—can view the history of each battery without any possibility of tampering. That assurance over data accuracy will enable Volkswagen to make much better decisions over the reuse, recycling, or disposal of the batteries while assuring consumers and regulators that, yes, the sustainability standards are upheld.
- BMW's Blockchain-enabled Battery Passport: BMW's Battery Passport program makes use of blockchain in developing a detailed digital record for each

EV battery. All relevant data related to battery materials, performance, and ownership history are on each passport, so accessible to every participant in the supply chain. It would ensure that BMW keeps updated records on battery usage and degradation/ownership by using blockchain with DT technology and ensuring data storage is safe. This system will make BMW environment friendly and instill consumers with confidence in tracing back its battery origin, usage, and recyclability. Blockchain also supports ethical sourcing and responsible asset management through such transparency.

These examples illustrate how blockchain can provide a trusted, tamper-proof track of data that makes compliance with the regulations easier and enhances sustainability in renewable energy. It fosters accountability and supports informed and responsible decision-making for each phase of reverse logistics by giving visibility into each asset's full life cycle.

10.2.2 Synergy of DTs with Blockchain

The cumulative state of DTs and blockchain gives the best approach to manage renewable energy assets in a sustainable, circular mode. Some such benefits can be observed along with blockchain and DTs as follows:

- Real-time Data Flow and Visibility: The integration of DTs and blockchain will validate status reverse logistics through asset condition data from each participant, and the decision-making for the EoL process shall be quicker and more precise.
- Better Decisions: It may be assured that the data flow between the DTs and blockchain systems is secure and transparent. For instance, an EV battery can report from a DT that it has hit the optimal point for recycling based on its performance metrics and market conditions (Bakhshi et al., 2024).
- Environmental Regulation Compliance: Since the tamper-proof blockchain maintains an accurate record for the proper use and recyclable asset history, adhering is easy to all regulatory compliance frameworks. A wind turbine operator can point out through auditable, clear evidence to the mandates of the sustainability (Wang et al., 2023).

This will also be used by companies for asset lifespans and optimizing resources in incorporating the DTs with blockchain in order to add up to the circular economy goals directly from minimized waste and enhanced reuse of valuable materials. Together, these innovations transform the reverse logistics function into secure, transparent, and efficient processes for the management of assets, thereby laying the foundation for a far stronger, more sustainable renewable energy sector.

10.2.3 Additional Considerations for DTs and Blockchain

Further in the chapter, other more profound means of value could be created for the reverse logistics of the renewable energy supply chains through DTs and blockchain.

DTs can utilize metrics beyond typical ones like energy output or degradation rates. For example, incorporating savings of carbon, rate of material recovery, or life-cycle energy consumption can provide a much more holistic view of what that asset implies in terms of sustainability (Askar et al., 2025). These metrics give both operating and strategic data leading to the achievement of greater environmental goals, and have a tendency to evoke more impactful decisions at organizational and policy levels.

Traceability, a strength of blockchain technology, indeed facilitates environmental compliance. However, up-to-date integration capability with regulatory update real-time in nature can provide the dynamic monitoring of compliance. As soon as the recycling rates or material recovery targets change, blockchain-enabled systems will be able to give automatic updates of compliance protocols and alert the stakeholders to risks of nonconformance. Dynamic will, therefore, unite static compliance reports and adaptive management for the benefit of providing significant value to stakeholders who wish to stay ahead of evolving regulations.

Then, for a more cross-sector interlinking, the blockchain and DT frameworks can work together. For instance, modeling material flows between and with other sectors could highlight renewable energy. Recycling of battery for consumer electronics or reusing turbine composites in construction would demonstrate such interconnections within the same circle sealed by blockchain and create an integrative, holistic circular economy at large. In turn, this would happen only if the framework could somehow also highlight such interconnections as one basis for the broader resource sustainability at play.

For DT and blockchain to maximally impact, these systems must address major issues related to interoperability and scalability (Alghamdi et al., 2024). Standardized data formats and protocols will be demanded in order to allow different systems across the supply chains of the world to integrate, but scalability will ensure that the framework can accommodate an increasing volume of assets with growing penetration of renewable energy.

10.3 Role of AI and IoT in Reverse Logistics Optimization

The complexity in reverse logistics in renewable energy systems requires much more advanced technologies, including AI and Internet of Things (IoT). AI and IoT can be combined with BEDT to provide better efficiency and predictability as well as sustainability in the operations within reverse logistics. This chapter discusses how AI and IoT contribute to optimized decision-making, improved asset management, and reduced inefficiencies in a circular economy for renewable energy.

10.3.1 Integration of AI into BEDT

Incorporating AI into the BEDT is an excellent upgrade in predictive prowess since it utilizes predictive analytics (Sjödin et al., 2023). In reverse logistics, AI can compute large datasets of DTs to predict the EoL of renewable energy assets. For instance, the wind turbines, solar panels, and EV batteries depreciate with time due to environmental factors, mechanical wearing, and usage patterns among other factors. The AI models could predict precisely when these resources are no longer efficient or safe to use, and stakeholders could prepare and take appropriate measures like scheduling repairs, replacements, or recycling processes (Spirito, 2024).

The benefits of reverse logistics include several key advantages from AI-driven predictive analytics:

- Optimized Resource Use: With AI-driven forecasting, one can plan resource usage more efficiently, as the recycling or refurbishment of assets happens only when they are really needed (Singh and Kaur, 2024).
- Reduced Downtime: The prediction of failures in renewable energy systems by AI ensures a reduction in unplanned downtime, which allows smoother operation (Afridi et al., 2022).
- Sustainability: AI contributes to the circular economy as it increases the lifespan of the asset and reduces waste due to accurate life-cycle predictions (Hariyani et al., 2024).

For example, consider the EoL stage of a vehicle's EV battery: AI can be used to analyze the battery's performance data against other peers in the same asset class and provide recommendations as to whether the battery needs to be refurbished for a new application, like energy storage, or is channeled directly to be recycled. Continuous learning through new data improves the efficiency over time of AI systems that apply reverse logistics.

This not only brings better decision-making for a single asset but enables creating a self-optimizing reverse logistics network. It continuously keeps tabs on the whole network of assets using AI, predicting bottlenecks, and adjusting logistics in real-time to cut down the inefficiencies within the systems (Skochynskyi and Скочинський, 2023). For example, an AI can dynamically reroute recycled materials to facilities that have the highest capacity or suggest transport routes that are the most energy-efficient and could reduce both cost and carbon footprint on operations. The main advantages of AI-based decision support are

- Superior decision-making: The AI process enormous amounts of data to suggest alternative with sustainable value, at low cost, and operate efficiently (Javaid et al., 2022).
- Self-optimizing networks: AI can build reverse logistics performance based on experience obtained from the current operations and eliminate waste and maximize recovery (Mourtzis et al., 2024).
- Real-time adaptability: AI can easily adapt itself to the changes in the market conditions, asset performance, or the needs of the regulatory authority and make reverse logistics responsive (Chen et al., 2024).

10.3.2 IoT-enabled Data Collection

The IoT is an enabler that enhances the accuracy and utility of DTs by providing real-time data on the health, performance, and environmental conditions of assets. The sensor-enabled IoT devices embedded in renewable energy systems such as wind turbines and solar panels or attached to EV batteries allow continuous monitoring through the entire life cycle of the asset, as in the example below:

- Wind Turbines: IoT sensors are incorporated with wind turbine blades and motors and measure critical parameters from the operating condition of such a turbine, including the speed of wind blade vibrations, structural fatigue temperature (Wang et al., 2022). Such sensors send live data of the operating conditions of the turbine so the wear or damage can be identified in advance. In such a case, it allows the DT of a turbine to perform predictive maintenance and optimal reverse logistics decisions for when the turbine or component is approaching EoL.
- Solar Panels: These are usually installed with sensors of IoT that monitor how efficiently they produce energy, and the degree of exposure to the environment, such as the growth of dust, temperature difference, etc. This is to ensure that the output of the panel is at all times monitored and determines when it should be cleaned, repaired, or substituted, thus optimizing their cycle of life (Gopal et al., 2020). Once these reach EoL, it is possible to consider refurbishment or recycling with the data generated.
- EV Batteries: IoT sensors in the EV batteries monitor the battery health parameters, which include temperature, voltage, and charging cycles. All of these are critical in tracking the degradation of the battery, ensuring that the battery is reused, repurposed, or recycled at the most optimal time (Sivaraman and Sharmeela, 2020).

IoT sensors collect real-time data on the conditions of assets, which improves the accuracy and reliability of DTs in better decision-making for reverse logistics. Such continuous streams of data assure stakeholders that the information used for decision-making can be devoid of unanticipated failures with proper resource planning.

With this regard, IoT-based data collection allows reverse logistics networks to be monitored and planned much better. According to accurate and real-time data gathered from IoT sensors, DTs can reproduce the actual state of renewable energy assets with much more accuracy and then make more informed and timely decisions about asset reuse, refurbishment, or recycling.

It improves the reverse logistics on the side of resource planning since through the real-time IoT data captured on the asset's availability and condition coming due for recycling, the recyclers and manufacturers have time to plan operations effectively to reduce delay and wastage.

IoT-enabled data is also helpful in complying with environmental regulations by providing an auditable record of the performance of assets and the EoL decisions. Renewable energy systems are thus sustainably managed and according to the legal requirements.

10.4 Case Studies and Applications

In recent years, various sectors such as EVs and renewable energy systems are also witnessing the promise of BEDT. The following case studies and applications are meant to represent real-world implementations of blockchain and DT technologies for optimizing reverse logistics and sustainability practices. This also centers on how the BEDT framework can be applied across industries because the applications thereof may be generalized into other renewable energy sectors such as wind and solar.

10.4.1 Industry Case Studies

1. **Volkswagen and IBM:** Volkswagen has taken fantastic steps in mainstreaming blockchain technology into its reverse logistics network (Santana, 2024). The company, with IBM, recently launched a pilot program that targets full traceability of EV batteries throughout their production, up to EoL disposal or recycling (IBM). It is therefore a system designed to take on the challenges of split data-that is, the battery's condition and ownership history must be tracked in as transparent and secure a fashion as possible.

   The Volkswagen-IBM project uses blockchain to form a decentralized ledger that tracks every transaction and event concerning an EV battery, thereby ensuring data is tamper-proof and accessible to all the stakeholders in the supply chain (Ledger Insights, 2019). This solution addresses critical problems in reverse logistics, such as the lack of real-time visibility into the usage and condition of a battery. This blockchain-based system, therefore, allows manufacturers and recyclers to make informed decisions on whether to recycle, reuse, or repurpose a battery.

This project provides the following important benefits:

- Better transparency of the life cycle of the battery will bring more informed decision-making into the recycling as well as reuse.
- Environmental regulation would be more compliant as blockchain creates a record from the time of battery production through to its eventual disposal.
- Optimization of reverse logistics operations as real-time data is available for the efficient scheduling of recycling or refurbishment processes.

This case study explores how the blockchain can be very suitable in conjunction with a reverse logistics framework to yield a more streamlined and efficient processes and higher sustainability. On the other hand, while the main aim of this project is its application to the principles of the EV battery, the same principle can easily be applied to renewable energy systems like the wind turbines and the solar panels.

2. **BMW: Digital Passports for EV Batteries:** BMW is another of the leaders in the automotive sector that is studying how blockchain technology could be applied to the EoL management of EV batteries (Berger et al., 2022; Hassini et al., 2024). BMW intends to deliver an extensive, secure data package that reflects the composition of a given battery, including material and usage, ownership history, and health through its "Battery Passport." In the digital passport approach, there is a guarantee that proper and real-time data can be accessed by everyone—from manufacturers and consumers to recyclers—and thus greater transparency and efficiency in reverse logistics.

   With BMW, for instance, blockchain offers a safe and decentralized system, where updating the digital passport for every battery becomes feasible in real-time as every battery passes through the value chain. The passport will contain the basic information necessary to decide the greenest course of action for the battery at its EoL—whether it should be recycled, reused, or repurposed. This system will also support regulatory compliance since every process will take into account the transparent and traceable nature of the battery-handling process.

   BMW's digital passport might, in the future, be combined with DT technology in such a way that real-time data fed from IoT sensors on the EV battery, to the blockchain system, makes this an even more powerful tool for optimization of reverse logistics (Rajesh et al., 2024). It would be able to automatically initiate recycling or refurbishment when particular conditions are met, for example, the level of battery degradation.

BMW is launching its project Cobalt for Development, which sources responsible cobalt from EV battery material supplies, in support of sustainability at a higher level, one focused on responsible sourcing of resources (Schmid, 2020). That helps promote recycling within the product life cycle to eventually help replace throwaway, waste-material economies with a circular economy.

3. **Renault's Digital Twin Approach:** Renault has been a leader in the adoption of DT technology for the management of the life cycle of EV batteries (Martinez et al., 2024; Stadtfeld et al., 2024). The company designed a system using DTs to monitor real-time health of its EV batteries – from manufacturing through usage and eventual recycling. The DT is the dynamic, data-driven model of the battery, constantly updating its performance, usage history, and environmental conditions.

   Renault's system also integrates AI predictive analytics in predicting battery failure and optimizing the recycling schedule (Hasan and Karim, 2024). It tracks real-time performance of batteries. In this manner, the battery is recycled only when it cannot be used any further through second-life applications, including stationary energy storage systems. The use of DTs also extends the lifespan of Renault batteries as it gives real-time information regarding how to further improve its efficiency and toughness.

   Although Renault applies DT mainly in EV batteries, there is every likelihood that its model could easily be stretched to apply in any other renewable energy systems. For example, one would easily use those same principles for monitoring how healthy wind turbine blades and solar panels are, hence maximizing EoL management optimization.

4. **Tesla's Battery-recycling Initiatives:** Tesla has been one of the leaders in the circular economy, especially following the management of EV batteries along sustainable lines. Their battery-recycling initiatives have created a closed-loop system to ensure an effective zero waste and maximum recovery of resources in the process (Naor, 2022). Such a system supports environmental sustainability while enhancing the efficiency of reverse logistics in the battery EoL management process.

   Tesla's battery-recycling model uses real-time data and high-tech tracking tools to monitor and optimize battery management from cradle to grave (Oroszlan, 2023). In this kind of model, there is a data-driven process of recycling each and every battery based on their usage history, health status, and readiness to reach its EoL. By collecting real-time data about battery health, Tesla can make better decisions on when that battery should be recycled, repurposed, or redirected toward secondary applications. This extraction system makes it possible for Tesla to extract lithium, cobalt, and nickel from old batteries, which reduces the demand for raw virgin materials and conserves resources.

To ensure traceability and transparency, Tesla embeds blockchain-based systems that track every action related to the recycling process (Jiang et al., 2022). Hence, from the time when it was first used to its last stages at the point of EoL will not be tampered with. In doing so, this process induces transparency among all stakeholders. Consequently, the same poses Tesla as consistent with environmental rules and regulations. Because this process will take place via a medium that produces an auditable data trail, the recycling efforts undertaken by Tesla serve as examples of the gold standard in terms of responsible sustainable practices related to battery management in electric vehicles.

5. **Other Initiatives in Renewable Energy:** Several sectors and initiatives within renewable energy are leading the front in actualizing circular economy practices, such as:
   - Ellen MacArthur Foundation's Circular Economy Initiative (Bocken and Short, 2020): This foundation engages in initiatives such as New Plastics Economy and Make Fashion Circular, which help develop models on resource efficiency and minimize waste. The foundation's CE100 programme nudges renewable energy companies toward regenerative design and recovery of EoL assets, thereby encouraging innovation in material reuse and reverse logistics.
   - First Solar Global Recycling Program (Chowdhury et al., 2020): The Global Recycling Program at First Solar can recover more than 90% of material from photovoltaic (PV) panels. Using real-time IoT data, the recycling process is optimized, reinforcing resource reuse and saving the environment.
   - Ambri Liquid Metal Battery Recycling (Yan et al., 2023): Ambri specializes in producing grid-scale liquid metal batteries and has made a recycling process for its products that ensures a minimum lifespan of 20 years for the batteries. These long lifetimes, along with recyclability, support sustainable circular economies in solutions offering battery storage.
   - Biomass Circularity Program by Veolia (Samberger, 2022): Organic waste is being converted to renewable energy through biomass plants in the network of Veolia in Europe. Forestry and agricultural by-products are being used as a source of fuel, followed by ash by-products from them being reused as soil additives. Thus, Veolia can minimize waste and bring full material reuse back to the ecosystem.
   - The UK-based firm, Drax Group, uses sustainable sources of biomass such as agricultural waste and wood residues to generate carbon-neutral energy (Di Gruttola, 2023). In addition, the firm intends to explore carbon capture

and storage (CCS) to further reduce emissions in shaping the circular economy with biomass.

- Hydro-Québec's Dam Equipment Reuse Program (Mohammadbeygy and Mohammadbeygy, 2024): For instance, Hydro-Québec gains significant component reuse and refurbishment from decommissioned dams, such as turbines and generators, which can be used for an extended period of time and mitigate waste. This is an example of circularity applied to the hydropower industry.
- Ormat Technologies Geothermal Brine Recycling (Ball, 2021): Ormat Technologies recycles geothermal brine, a by-product of geothermal electricity, for the extraction of valuable minerals such as lithium and silica. It serves to decrease the environmental footprint, offers new sources of revenue, and enhances resource efficiency in geothermal energy.

These examples represent different applications of the circular economy across renewable energy sectors, thereby showcasing appropriate approaches for sustainable resource management, waste reduction, and extended asset life cycles.

10.4.2 Real-world Application to Renewable Energy

The foregoing case studies from the automobile sector demonstrate a massive application area of blockchain and DT for reverse logistics of batteries used in EVs. This, however, doesn't restrict their applicability just to the automotive sector, as blockchain and DT technologies could easily be generalized and extended into other renewable energy plants as well. The systems mentioned here are wind turbines and solar panels.

1. **BEDT Framework in Wind Farm Decommissioning:** Perhaps the greatest problem is that of reverse logistics in wind farms. It is going to be several years before the first-generation wind turbines are slated for retirement, which creates an enormous problem with the blades themselves. Because fiberglass composites have never been recycled and are used in making wind turbine blades, they obviously cannot be recycled as of yet (Bulińska et al., 2024). But via BEDT, the operators can monitor in real-time the wind farm's turbines and thus are ideally positioned to determine precisely when and how the blades will be recycled.

   As an example, leading wind turbine manufacturer Siemens Gamesa has been seeking to make its blades more recyclable (Vella, 2022). By combining DT technology with blockchain, Siemens Gamesa could develop a system in which each blade had a digital twin that could track usage, wear, and environmental exposure throughout the life cycle. This will help predict when a particular blade can be replaced, thus getting the recycling processes to take place at the opportune best time.

2. **Solar Panel Recycling and the BEDT Framework:** The other area that poses big challenges to reverse logistics is the solar energy systems. Though the sun continues to gain appreciation in most regions of the world, it still calls for the development of some strategies relating to EoL management. Usually, these solar panels last for 25–30 years, and then they need recycling or refurbishment and ultimately are disposed of.

   The Solar Energy Industries Association has begun a recycling program for solar panels in the United States, but it is still very fractured and does not have an overall system that tracks the health of individual panels (Curtis et al., 2021). Using the BEDT framework, the manufacturers and operators of the solar panels can monitor each panel's real-time performance using IoT sensors. This information can then be stored on a blockchain, so it is a safe and tamper-proof record of the condition of the panel, the materials, and usage history.

   Besides all of the above, the BEDT framework can be extended to energy storage units and hydropower equipment, among other renewable energy systems, which would make it a highly versatile tool for optimizing reverse logistics in many different types of industries.

10.5 Smoothing the Transition to Renewable Energy

The future transition of the world from fossil energy toward renewable energy entails more than a shift toward clean source use; the process would also need sustainability and be considered as going in the way of a well-rounded life cycle for renewable energy-based assets. With further expansion taking place in many sectors in terms of wind, solar energy, or EVs, one would surely face serious challenges toward responsible closure and end when the lifetime ends. This section will detail the role BEDT can play in such transitions by making reverse logistics possible for an enhanced-efficiency system to support circular economy principles, and environmental regulation compliance.

10.5.1 BEDT's Role in Transitioning to Renewable Energy

BEDT bring an innovative way to address a few of the most vital issues at the renewable energy front as they optimize the reverse logistics process. Usually, old reverse logistics networks are unable to provide required transparency, efficiency, and security to be maintained for responsible EoL management of renewable assets. BEDT gives renewable energy sectors a completely new frontier of operational intelligence, therefore, reducing wastes and achieving better life-cycle management (Roumeliotis et al., 2024).

Utilizing BEDT technology, one can determine the real-time condition data of usage and environmental impact on assets like EV batteries, wind turbines, and solar panels. One can then make well-informed decisions about recycling, reusing, or repurposing at the appropriate times. For instance, for EVs, BEDT improves the reverse logistics process because it can monitor in real-time the health and condition of the battery for its recycle or repurpose at just the right time. Such strategic interventions significantly reduce waste, thereby averting early disposal and creating a ripple of great environmental results in renewable energy supply chains.

BEDT can address the widespread issues of data fragmentation and obsolescence so that the management of the network of renewable energy assets at EoL becomes less fragmented and more transparent (Li and Kassem, 2021). Improved traceability for the origin, composition, and condition of renewable energy assets helps stakeholders continuously monitor such traces. The traceability would not only make reverse logistics operations easier but would also recover and reuse the materials found within these assets in a much better way to create a sustainable supply chain in the renewable energy sectors.

10.5.2 Contributions to the Circular Economy and Sustainability

One of the defining features of BEDT is promoting a circular economy: extending the life cycle of renewable energy assets and reducing resource extraction. In the circular economy, resources are continuously being reused, remanufactured, or recycled and significantly reduce raw material extraction and waste generation. BEDT supports this model through comprehensive, real-time data that guides optimal asset management from maintenance to EoL recycling or repurposing. Because each DT tracks the life-cycle data of its physical counterpart, it empowers companies to make informed decisions about reusing or recycling components, thereby directly contributing to resource conservation and waste reduction.

Besides, BEDT frameworks allow companies to operate under an ever-changing scope of environmental and waste management regulations. Most regions will mandate companies to manage their EoL processes, ensuring renewable energy assets do not damage the environment post their use. The blockchain-based feature of BEDT makes possible an auditable record, tamper-proof account for every asset's life cycle ranging from manufacturing to disposal. With this immutable record, the company can meet regulation much more effectively, diminishing risks of noncompliance with the regulations and showing responsibility to consumers and all the stakeholders.

Beyond legal compliance, the impacts of BEDT extend the impacts of sustainability. BEDT will support source recovery efficiently with minimum environmental effects in line with the global sustainability goals and United Nations SDGs.

Companies contribute to the sustainable energy transition through incorporation of BEDT in reverse logistics that reduces carbon footprints and builds long-term ecological resilience.

10.6 Conclusion and Way Forward

The BEDT framework is, therefore, promising in the application to enhance reverse logistics operations in renewable energy so as to provide higher transparency, efficiency, and compliance in the life cycle of assets, such as EV batteries, wind turbines, and solar panels. The integration of digital twins with blockchain technology allows BEDT to develop an auditable, tamper-proof record of asset conditions and ownership while simultaneously supporting real-time monitoring and predictive maintenance, thereby aligning asset management with circular economy goals. This framework, in fact, provides a basis for renewable energy systems to evolve from linear to circular models of economy through the avoidance of waste and maximized asset life cycles through reuse, refurbishment, and recycling.

Future research should focus on further expanding the capabilities of BEDT with advanced machine learning algorithms that can refine predictive analytics to inform decision-makers of whether or when to recycle or repurpose assets. In addition, developments in universally accepted smart contracts can automate any of the key processes in reverse logistics and especially trigger recycling action when thresholds of their degradation are attained. Moreover, it would allow bringing scalable IoT sensor networks with much more accurate real-time data collection, making the digital twins as exact as possible. It is also necessary to overcome interoperability challenges between IoT devices and blockchain platforms while building a unified responsive data ecosystem within the BEDT framework.

Moreover, further development of the application of BEDT beyond renewable energy sectors currently embraced is still another valuable direction. For instance, BEDT can also be implemented in other sustainable industries like hydropower and bioenergy, where long-life equipment and material life-cycle management support a more sustainable development. Furthermore, the research into carbon credit integration in BEDT can further motivate firms to continue their sustainable reverse logistics with additional sources of revenue as well as enhance them to carry greater environmental responsibilities. Its implications for the circular economy may also be expanded to the implementation of BEDT in the management of postconsumer waste by industries with a high demand on the use of rare earth materials.

Building cross-sector collaborations and partnerships will be important in furthering standards and regulatory frameworks for BEDT and in creating the best practices of industry. Accelerating engagements with government, academia, and industry leaders will drive the development and adoption of BEDT. It stands to move into the providing of innovative solutions for solving the thorny issues in the supply chain. In general, the BEDT model is a model looking forward toward identifying new ways of practicing sustainability and economic growth but, most importantly, dealing with reverse logistics in a way that negates inefficiencies. All these developments will contribute significantly toward making a more resilient and sustainable global energy landscape possible.

Bibliography

E Acosta Llano, P Hurmelinna-Laukkanen, and L Haapanen. Blockchain for the circular economy, implications for public governance. *International Journal of Public Sector Management*, 38(1): 30–52, 2024. https://doi.org/10.1108/IJPSM-12-2023-0365/FULL/PDF.

YS Afridi, K Ahmad, and L Hassan. Artificial intelligence based prognostic maintenance of renewable energy systems: A review of techniques, challenges, and future research directions. *International Journal of Energy Research*, 46(15): 21619–21642, 2022. https://doi.org/10.1002/ER.7100.

TA Alghamdi, R Khalid, and N Javaid. A survey of blockchain based systems: scalability issues and solutions, applications and future challenges. *IEEE Access*, 12: 79626–79651, 2024. https://doi.org/10.1109/ACCESS.2024.3408868.

A Ali, MT Islam, S Rehman, SA Qadir, M Shahid, MW Khan, MH Zahir, A Islam, and M Khalid. Solar photovoltaic module end-of-life waste management regulations: International practices and implications for the Kingdom of Saudi Arabia. *Sustainability*, 16(16): 7215, 2024. https://doi.org/10.3390/SU16167215.

R Askar, F Karaca, A Salles, A Lukyanenko, GC Cervantes Puma, V Tavares, A Khaidarova, A Nadaždi, R Pineda-Martos, JM Díaz-Cabrera, et al. Driving the built environment twin transition: Synergising circular economy and digital tools. In *Circular Economy Design and Management in the Built Environment: A Critical Review of the State of the Art*, pp. 459–505. Cham, Springer Nature, 2025. https://doi.org/10.1007/978-3-031-73490-8_17.

PJ Ball. A review of geothermal technologies and their role in reducing greenhouse gas emissions in the USA. *Journal of Energy Resources Technology, Transactions of the ASME*, 143(1): 010903, 2021. https://doi.org/10.1115/1.4048187/1086526.

S Bakhshi, A Ghaffarianhoseini, A Ghaffarianhoseini, M Najafi, F Rahimian, C Park, and D Lee. Digital twin applications for overcoming construction supply chain challenges. *Automation in Construction*, 167: 105679, 2024. https://doi.org/10.1016/J.AUTCON.2024.105679.

K Berger, JP Schöggl, and RJ Baumgartner. Digital battery passports to enable circular and sustainable value chains: Conceptualization and use cases. *Journal of Cleaner Production*, 353: 131492, 2022. https://doi.org/10.1016/J.JCLEPRO.2022.131492.

NMP Bocken and SW Short. Transforming business models: Towards a sufficiency-based circular economy. In *Handbook of the Circular Economy*, pp. 250–265. Edward Elgar Publishing, 2020. https://doi.org/10.4337/9781788972727.00028.

Y Bouzarour-Amokrane, A Tchangani, and F Peres. Decision evaluation process in end-of-life systems management. *Journal of Manufacturing Systems*, 37: 715–728, 2015. https://doi.org/10.1016/J.JMSY.2015.03.007.

R Brandín and S Abrishami. Information traceability platforms for asset data lifecycle: Blockchain-based technologies. *Smart and Sustainable Built Environment*, 10(3): 364–386, 2021. https://doi.org/10.1108/SASBE-03-2021-0042/FULL/XML.

S Bulińska, A Sujak, and M Pyzalski. From waste to renewables: Challenges and opportunities in recycling glass fibre composite products from wind turbine blades for sustainable cement production. *Sustainability*, 16(12): 5150, 2024. https://doi.org/10.3390/SU16125150.

AS Butt, I Ali, and K Govindan. The role of reverse logistics in a circular economy for achieving sustainable development goals: A multiple case study of retail firms. *Production Planning & Control*, 35(12): 1490–1502, 2024. https://doi.org/10.1080/09537287.2023.2197851.

W Chen, Y Men, N Fuster, C Osorio, and AA Juan. Artificial intelligence in logistics optimization with sustainable criteria: A review. *Sustainability*, 16(21): 9145, 2024. https://doi.org/10.3390/SU16219145.

M Chetan, S Yao, and DT Griffith. Multi-fidelity digital twin structural model for a sub-scale downwind wind turbine rotor blade. *Wind Energy*, 24(12): 1368–1387, 2021. https://doi.org/10.1002/WE.2636.

MS Chowdhury, KS Rahman, T Chowdhury, N Nuthammachot, K Techato, M Akhtaruzzaman, SK Tiong, K Sopian, and N Amin. An overview of solar photovoltaic panels' end-of-life material recycling. *Energy Strategy Reviews*, 27: 100431, 2020. https://doi.org/10.1016/J.ESR.2019.100431.

TL Curtis, H Buchanan, G Heath, L Smith, and S Shaw. Solar photovoltaic module recycling: A survey of U.S. policies and initiatives. National Renewable Energy Laboratory, 2021. https://doi.org/10.2172/1774839.

M Dietz and G Pernul. Digital twin: Empowering enterprises towards a system-of-systems approach. *Business and Information Systems Engineering*, 62(2): 179–184, 2020. https://doi.org/10.1007/S12599-019-00624-0/FIGURES/2.

F Di Gruttola. Thermo-Catalytic Reforming of waste biomass for alternative fuels production in a framework of decarbonization, 2023. Accessed: Nov. 14, 2024. [Online]. Available: https://iris.uniroma1.it/handle/11573/1672716.

ML Di Silvestre, S Favuzza, E Riva Sanseverino, and G Zizzo. How decarbonization, digitalization and decentralization are changing key power infrastructures. *Renewable and Sustainable Energy Reviews*, 93: 483–498, 2018. https://doi.org/10.1016/J.RSER.2018.05.068.

OT Eleftheriou and C-N Anagnostopoulos. Digital twins: A brief overview of applications, challenges and enabling technologies in the last decade. *Digital Twin*, 2: 2, 2022. https://doi.org/10.12688/DIGITALTWIN.17581.1.

X Fan and Y Li. Energy management of renewable based power grids using artificial intelligence: Digital twin of renewables. *Solar Energy*, 262: 111867, 2023. https://doi.org/10.1016/J.SOLENER.2023.111867.

Z Gao. The coming tide of wind turbine blades retirement: Threats and treatment measures. *Highlights in Science, Engineering and Technology*, 29: 105–112, 2023. https://doi.org/10.54097/HSET.V29I.4487.

D Gautam and N Bolia. Developing an incentive-based model for efficient product recovery and reverse logistics. *Business Strategy and the Environment*, 33(8): 7972–7989, 2024. https://doi.org/10.1002/BSE.3906.

M Gopal, T Chandra Prakash, N Venkata Ramakrishna, and BP Yadav. IoT based solar power monitoring system. In *IOP Conference Series: Materials Science and Engineering*, volume 981, number 3, page 032037. IOP Publishing, 2020. https://doi.org/10.1088/1757-899X/981/3/032037.

O Hakimi, H Liu, O Abudayyeh, A Houshyar, M Almatared, and A Alhawiti. Data fusion for smart civil infrastructure management: A conceptual digital twin framework. *Buildings*, 13(11): 2725, 2023. https://doi.org/10.3390/BUILDINGS13112725.

D Hariyani, P Hariyani, S Mishra, and M Kumar Sharma. Leveraging digital technologies for advancing circular economy practices and enhancing life cycle analysis: A systematic literature review. *Waste Management Bulletin*, 2(3): 69–83, 2024. https://doi.org/10.1016/J.WMB.2024.06.007.

F Hasan and S Karim. Role of digitalization and digital capabilities to implement second-life business models for electric vehicle batteries. Mälardalens University, 2024, Accessed: Nov. 14, 2024. [Online]. Available: https://urn.kb.se/resolve?urn=urn:nbn:se:mdh:diva-67874.

M Hassini, E Redondo-Iglesias, and P Venet. Battery passports for second-life batteries: An experimental assessment of suitability for mobile applications. *Batteries*, 10(5): 153, 2024. https://doi.org/10.3390/BATTERIES10050153.

IBM. FAW-Volkswagen, 2022. Accessed: Nov. 13, 2024. [Online]. Available: https://www.ibm.com/case-studies/faw-volkswagen.

FB Ismail, H Al-Faiz, H Hasini, A Al-Bazi, and HA Kazem. A comprehensive review of the dynamic applications of the digital twin technology across diverse energy sectors. *Energy Strategy Reviews*, 52: 101334, 2024. https://doi.org/10.1016/J.ESR.2024.101334.

M Javaid, A Haleem, RP Singh, R Suman, and ES Gonzalez. Understanding the adoption of Industry 4.0 technologies in improving environmental sustainability. *Sustainable Operations and Computers*, 3: 203–217, 2022. https://doi.org/10.1016/J.SUSOC.2022.01.008.

L Jiang, Y Gu, W Yu, and J Dai. Blockchain-based life cycle assessment system for ESG reporting. *SSRN Electronic Journal*, 2022. https://doi.org/10.2139/SSRN.4121907.

SS Kamble, A Gunasekaran, H Parekh, V Mani, A Belhadi, and R Sharma. Digital twin for sustainable manufacturing supply chains: Current trends, future perspectives, and an implementation framework. *Technological Forecasting and Social Change*, 176: 121448, 2022. https://doi.org/10.1016/J.TECHFORE.2021.121448.

Ledger Insights. Volvo joins VW, Ford, IBM for blockchain cobalt traceability. *Ledger Insights - Blockchain for Enterprise*, 2019. Accessed: Nov. 14, 2024. [Online]. Available: https://www.ledgerinsights.com/volvo-vw-ford-ibm-blockchain-cobalt-traceability-responsible-minerals-rsbn/.

J Li and M Kassem. Applications of distributed ledger technology (DLT) and Blockchain-enabled smart contracts in construction. *Automation in Construction*, 132: 103955, 2021. https://doi.org/10.1016/J.AUTCON.2021.103955.

J Martinez, I Bouhali, L Palladino, V Idasiak, F Kratz, JY Choley, and F Mhenni. Accelerating digital transformation through MBSE, multi-physics simulation and digital twin in industry. *INCOSE International Symposium*, 34(1): 691–715, 2024. https://doi.org/10.1002/IIS2.13171.

T Mohammadbeygy and T Mohammadbeygy. The wheels of progress: Challenging quebec's sustainable mobility plan from an ecological economics perspective, McGill University, 2024. Accessed: Nov. 14, 2024. [Online]. Available: https://escholarship.mcgill.ca/concern/theses/765376486?locale=en.

D Mourtzis, N Panopoulos, P Stavropoulos, and N Papakostas. Artificial intelligence for production management and control towards mass personalization of global networks. In *CIRP Novel Topics in Production Engineering: Volume 1*, pp. 267–312. Springer Nature, 2024. https://doi.org/10.1007/978-3-031-54034-9_8.

M Naor. Tesla's circular economy strategy to recycle, reduce, reuse, repurpose and recover batteries. In *Recycling Strategy and Challenges Associated with Waste Management Towards Sustaining the World*. IntechOpen, 2022. https://doi.org/10.5772/INTECHOPEN.107256.

AM Omer. Energy, environment and sustainable development. *Renewable and Sustainable Energy Reviews*, 12:9:2265–2300, 2008. https://doi.org/10.1016/J.RSER.2007.05.001.

D Oroszlan. High school understanding of the cradle-to-grave process of electric vehicles. *Journal of Student Research*, 12(4), 2023. https://doi.org/10.47611/jsrhs.v12i4.5868.

K Psara, C Papadimitriou, M Efstratiadi, S Tsakanikas, P Papadopoulos, and P Tobin. European energy regulatory, socioeconomic, and organizational aspects: An analysis of barriers related to data-driven services across electricity sectors. *Energies*, 15(6): 2197, 2022. https://doi.org/10.3390/EN15062197.

PK Rajesh, T Soundarya, and KV Jithin. Driving sustainability-the role of digital twin in enhancing battery performance for electric vehicles. *Journal of Power Sources*, 604: 234464, 2024. https://doi.org/10.1016/J.JPOWSOUR.2024.234464.

C Roumeliotis, M Dasygenis, V Lazaridis, and M Dossis. Blockchain and digital twins in smart industry 4.0: The use case of supply chain-a review of integration techniques and applications. *Designs*, 8(6): 105, 2024. https://doi.org/10.3390/DESIGNS8060105.

C Samberger. The role of water circularity in the food-water-energy nexus and climate change mitigation. *Energy Nexus*, 6: 100061, 2022. https://doi.org/10.1016/J.NEXUS.2022.100061.

RA Santana. Improving visibility to foster more informed decision making: The case of containers reverse logistics at Volkswagen AutoEuropa. Master's thesis, ISCTE-Instituto Universitario de Lisboa, 2024. Accessed: Nov. 13, 2024. [Online]. Available: https://repositorio.iscte-iul.pt/handle/10071/32175.

M Schmid. Challenges to the European automotive industry in securing critical raw materials for electric mobility: The case of rare earths. *Mineralogical Magazine*, 84(1): 5–17, 2020. https://doi.org/10.1180/MGM.2020.9.

M Singh and G Kaur. Case studies on AI-driven innovations in renewable energy, waste management, and resource conservation. In *Maintaining a Sustainable World in the Nexus of Environmental Science and AI*, pp. 455–484. IGI Global, 2024. https://services.igi-global.com/resolvedoi/resolve.aspx?doi=10.4018/979-8-3693-6336-2.ch018.

P Sivaraman and C Sharmeela. IoT-based battery management system for hybrid electric vehicle. In *Artificial Intelligent Techniques for Electric and Hybrid Electric Vehicles*, pp. 1–16. Wiley AI, 2020. https://doi.org/10.1002/9781119682035.CH1.

D Sjödin, V Parida, and M Kohtamäki. Artificial intelligence enabling circular business model innovation in digital servitization: Conceptualizing dynamic capabilities, AI capacities, business models and effects. *Technological Forecasting and Social Change*, 197: 122903, 2023. https://doi.org/10.1016/J.TECHFORE.2023.122903.

B Skochynskyi and БД Скочинський. Methodology for optimizing the logistics of product purchases in the war and post-war period, p. 107. National Aviation University, 2023. Accessed: Nov. 13, 2024. [Online]. Available: https://er.nau.edu.ua/handle/NAU/63361.

J Song, G He, J Wang, and P Zhang. Shaping future low-carbon energy and transportation systems: Digital technologies and applications. *iEnergy*, 1(3): 285–305, 2022. https://doi.org/10.23919/IEN.2022.0040.

C Spirito. Artificial Intelligence applications in Reverse Logistics, how technology could improve return and waste management creating value. Doctoral dissertation, Politecnico di Torino, 2024.

S Srivastava, Mohit, A Kumar, SK Jha, P Dixit, and S Prakash. Event-driven data alteration detection using block-chain. *Security and Privacy*, 4(2): e146, 2021. https://doi.org/10.1002/SPY2.146.

GM Stadtfeld, R Lienemann, and T Gruchmann. An analysis of digital twin technologies enhancing supply chain viability: Empirical evidence from multiple cases. *Production Planning and Control*, 1–17, 2024. https://doi.org/10.1080/09537287.2024.2415418.

P Tavakoli, I Yitmen, H Sadri, and A Taheri. Blockchain-based digital twin data provenance for predictive asset management in building facilities. *Smart and Sustainable Built Environment*, 13(1): 4–21, 2024. https://doi.org/10.1108/SASBE-07-2023-0169/FULL/PDF.

H Vella. Turbines to catch a second wind [wind power - recycling]. *Engineering & Technology*, 17(10): 38–41, 2022. https://doi.org/10.1049/ET.2022.1013.

E Walker and M Casey. How long do solar panels last? | EnergySage, 2024. Accessed: Aug. 14, 2025. [Online]. Available: https://www.energysage.com/solar/how-long-do-solar-panels-last/.

B Wang, Z Lin, M Wang, F Wang, P Xiangli, and Z Li. Applying blockchain technology to ensure compliance with sustainability standards in the PPE multi-tier supply chain. *International Journal of Production Research*, 61(14): 4934–4950, 2023. https://doi.org/10.1080/00207543.2022.2025944.

W Wang, Y Xue, C He, and Y Zhao. Review of the typical damage and damage-detection methods of large wind turbine blades. *Energies*, 15(15): 5672, 2022. https://doi.org/10.3390/EN15155672.

M Wilson and S Goffnett. Reverse logistics: Understanding end-of-life product management. *Business Horizons*, 65(5): 643–655, 2022. https://doi.org/10.1016/J.BUSHOR.2021.10.005.

S Yan, H Li, W Chen, H Zhou, X Zhou, W Zhang, Z Li, Y He, K Wang, and K Jiang. Full-lifetime recycling and reutilization of key materials of low-cost and sustainable liquid metal batteries. *ACS Sustainable Chemistry & Engineering*, 11(48): 17038–17045, 2023. https://doi.org/10.1021/ACSSUSCHEMENG.3C05141/SUPPL_FILE/SC3C05141_SI_001.PDF.

C Yu, M Moslehpour, TK Tran, LM Trung, JP Ou, and NH Tien. Impact of non-renewable energy and natural resources on economic recovery: Empirical evidence from selected developing economies. *Resources Policy*, 80: 103221, 2023. https://doi.org/10.1016/J.RESOURPOL.2022.103221.

Y Zhao, O Pohl, AI Bhatt, GE Collis, PJ Mahon, T Rüther, and AF Hollenkamp. A review on battery market trends, second-life reuse, and recycling. *Sustainable Chemistry*, 2(1): 167–205, 2021. https://doi.org/10.3390/SUSCHEM2010011.

11

Pathways to AI Adoption in Offshore Wind Energy Operations and Maintenance

Callum Rothon[1,2] and Nina Dethlefs[1]*

[1] *Language and Data Research Group, Loughborough University, Loughborough, Leicestershire, UK*
[2] *Aura Centre for Doctoral Training, Energy & Environment Institute, University of Hull, Hull, East Riding of Yorkshire, UK*

*Corresponding author: Callum Rothon; cocr2@lunet.lboro.ac.uk

11.1 Introduction

In recent years, the offshore wind (OSW) energy and artificial intelligence (AI) industries have grown in tandem. AI tools have become commonplace in everyday life, with LLMs being adopted virtually overnight since the launch of ChatGPT in 2022 (OpenAI, 2022). Meanwhile, the OSW industry has grown worldwide, with the North Sea and Asia leading in manufacturing and adoption. In the United Kingdom, OSW provided 49 TWh in 2023, enough to power 50% of homes, representing 17% of total electricity generation (The Crown Estate, 2024). The United Kingdom has committed to installing 300 GW of OSW power generation by 2050, more than doubling current capacity.

As the OSW industry has matured, the volume of data being gathered from assets has increased, with current wind turbines (WTs) including monitoring of wind speed, temperatures, power output, and vibrations among their SCADA outputs. The range and volume of data available allow the behavior of assets to be monitored at a component level (e.g. a gearbox), an asset level (e.g. a turbine), and at a fleet level (e.g. a whole wind farm). The drive toward Smart O&M in the OSW industry has also led to changes in inspection methods, with the aim of reducing offshore hours for staff, and reducing downtime and lost generation while WTs

Renewable Energy Transition with Artificial Intelligence: Challenge-driven Solutions, First Edition. Nina Dethlefs and Joyjit Chatterjee. Published 2026 by John Wiley & Sons, Inc.

are being inspected and maintained. Inspections have moved from manual rope-access inspections to the use of remote cameras and UAVs with a range of imaging technologies (BVGAssociates, 2019).

This large volume and variety of data being generated makes the OSW industry a prime candidate for AI integration, with many operators already adopting AI tools, and with the industry poised to further adopt AI in a range of settings. However, while AI tools have developed rapidly and shown promising performance, they are a new and unproven technology, with drawbacks and limitations alongside their widely publicized strengths. If AI is to reach its full potential in industry, concerns around dependability and the sustainability of their use must be allayed, or the applications of AI tools must be carefully considered and justified.

In order to reach current sustainability targets, the OSW industry is moving toward the use of AI solutions in a range of applications. In the Offshore Wind Industrial Growth Plan 2024, Renewable UK (2024) recommends the integration of machine learning (ML) algorithms for Environmental Surveys and O&M by 2025 and 2032, respectively, highlighting the rapid adoption expected.

In this chapter, we consider how various factors holding back the adoption of AI in the OSW industry may be tackled, considering the stakeholders involved in bringing AI methods from research to application in OSW O&M. We consider the business and environmental cases for AI adoption, and the technical capabilities and limitations of current AI methods, and recommend pathways for AI adoption, matching recent developments in AI to the current challenges. We also consider the softer influences of nonexpert opinion and communication around AI, and how these factors may shape the future of AI in the OSW industry.

11.1.1 Current Applications of AI in OSW O&M

AI tools have already been applied to a range of problems in the OSW industry (Rashid et al., 2024; Hamdan et al., 2024). Dörterler et al. (2024) review applications of AI in wind energy, and find that AI solutions have been adopted in the OSW due to their deep analysis and prediction capabilities. The citation densities of AI methods for various tasks are shown in Figure 11.1.

In a keyword search of 1456 publications archived on Web of Science between 1992 and 2023, Dörterler et al. (2024) found that the most popular applications for AI in the OSW are forecasting, optimization, fault detection, condition monitoring, and decision-making. Forecasting alone accounts for over a quarter of all citations, showing focus on AI models as predictive tools in academia. Similar results are shown when identifying the distribution of applications in industry, with forecasting and optimization at 30% and 22% of applications, respectively, showing agreement between academia and the OSW industry.

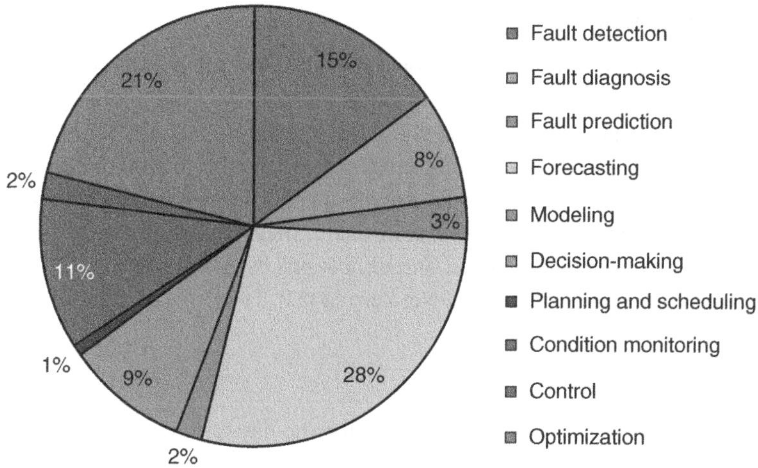

Figure 11.1 Citation density for AI applications in the OSW industry (Dörterler et al. [2024] / John Wiley & Sons).

Time-series forecasting (TSF) models have been applied to predicting wind conditions, and the resulting power generation. Accurate prediction of future generation is key to optimizing generation, grid integration, availability, and profitability of OSW energy. Modern forecasting models, including transformers and LSTMs, are able to learn long-term dependencies in data, and predict multiple time steps ahead with high accuracy, making them highly suitable to these tasks.

AI models have also been applied to condition monitoring, including the early detection and prediction of faults, using a range of techniques and data sources. The continuous monitoring of SCADA data allows faults to be detected earlier than inspections, with classification models capable of identifying anomalies and latent faults before a failure occurs. State-of-the-art computer vision models are capable of near expert-level performance of 95% in damage detection and classification tasks on UAV-gathered images of turbines (Iyer et al., 2022), supplementing human expertise during inspections.

Time-series models have been implemented that are capable of predicting future failures based on current trends. ORE Catapult identifies improved reliability, reduced downtime, revenue savings, and improved supply chain coordination as benefits of adopting AI tools in O&M in the OSW industry (ORE Catapult, 2024).

The adoption of these technologies has been driven by the need for rapid development to meet demand for renewable energy, while making OSW generation profitable and reducing the levelized cost of energy (LCoE) to be favorable compared to fossil fuels and other renewables.

11.1.2 Defining Stakeholders

Before we can discuss pathways to AI adoption in different areas of the OSW, we must consider the various stakeholders involved in AI in the OSW industry. The stakeholder groups that we discuss are wind farm operators, in-house O&M divisions and external contractors, academia and research, and the general public. For these stakeholders, we consider their overarching goals, their incentives for and against the adoption of AI tools, the level of influence they have, and how they may interact with other stakeholders based on publications by BVGAssociates (2019), The Department for Energy Security & Net Zero (2023), The Crown Estate (2024), and Renewable UK (2024).

Operators

Operators are companies that are responsible for the day-to-day running of the wind farm, including optimizing energy output, monitoring performance, and conducting inspections and maintenance. Operators will be driven by financial concerns, aiming to reduce costs, maximize profits and shareholder value, and win future contracts and bids.

We consider that operators have high power over the adoption of AI, and have very high interest in the costs and benefits. Generally, operators will have final say over the adoption of AI tools, either through their own policies and in-house O&M procedures, or through the equipment that they purchase and external O&M contractors that they hire.

O&M

O&M on a wind farm includes a range of activities, including monitoring and inspection, routine servicing, repairs, and other maintenance activities throughout the lifetime of the wind farm. This work is carried out by a mixture of staff from operators' internal O&M divisions, and by external contractors, depending on the wind farm and task.

From an O&M perspective, the main drivers of uptake of AI technologies are feasibility and practicality, where solutions that are safer, simpler, and more effective than current methods are more likely to be adopted. As with operators, there will be the requirement for technologies to increase profitability over existing methods.

O&M divisions and contractors will influence the adoption of AI by using and developing tools that improve efficiency, balancing profitability and practical concerns, although they are ultimately bound by the policies of operators.

Academia and Research

Academics and research staff will influence the adoption of AI in the OSW industry by developing methods and tools, which the industry will take up if they see a benefit. Some academics will be more focused on pure research and publications, while others may seek to commercialize their research through SMEs or through collaboration with larger firms including operators. Academics and researchers in the field of AI will have a high interest in the development of new methods and in promoting progress in the field. Leading academics and thought leaders will have high influence over nonexpert opinion on AI, which will influence how technologies are adopted. AI researchers working in the OSW industry will naturally have a high interest in the adoption of AI tools, and will have greater influence on applications than academics in the broader field of AI given their close links to industry.

Public

The general public and shareholders are the least influential and least engaged stakeholders at an individual level. Provided that energy costs do not increase too rapidly, and shareholder value increases, these stakeholders are satisfied. Current public opinion is gradually shifting away from fossil fuels and toward green energy, and AI is being adopted en masse, although both with some resistance. The overall desires of the public will be balanced somewhere between their economic needs for affordable and reliable energy, and their opinions on wind power and AI.

As a population, the general public can exert influence by voting for policies on renewable energy or by lobbying for or against renewables in their local areas, but this is limited. Shareholders in operators and utility companies can be considered similarly, although with higher interest in share prices and more direct influence over corporate decisions if moving in a block. Although the influence of the general public is low on an individual level, the impacts of public opinion on government policy can be significant, considering the importance of agreedupon strike prices and contracts for difference to the profitability of OSW generation.

11.2 Technical Challenges and Solutions

11.2.1 Setup and Running Costs of AI

The OSW industry deals in massive sums of money, with modern turbines costing upwards of £10000 000, and a single rotation of a 14-MW turbine capable of powering a family home for two days. The setup costs of AI tools can be high, with leading models requiring significant computing power to train and run, significant amounts of data to train on, and using large amounts of energy. The two main routes to return on investment from adopting AI tools are by reducing costs and maximizing profitability of generation over the lifetime of the wind farm.

Hardware

A key factor in the rise of AI technologies has been the increase in performance and reduction in cost of GPUs, which are used to accelerate training and running of deep learning (DL) programs. In recent years, GPUs have accelerated faster than CPUs due to a combination of improvements in architecture, interconnects, memory technology, and algorithms (Perry, 2018). This has led to GPUs becoming dominant in AI research. While large models require clusters of powerful GPUs arranged as high-performance computers, consumer-level GPUs can also be used, reducing barriers to entry for AI training and use.

Software

Ready-to-use AI solutions are available for a range of use cases, but these can be expensive, especially at a large scale. Many AI libraries, including TensorFlow (Abadi et al., 2015) and PyTorch (Paszke et al., 2019), are open source, so are free to use. However, some expertise is still required to implement, train, and use models based on these libraries, so staffing costs must be considered.

Data

When training AI models, large amounts of data are required to maximize model performance, with more varied data giving a more generalizable model. In highly specific use cases, smaller amounts of data may suffice, but in most cases the more data the better the training. In cases where data is not publicly available, data must be gathered. Depending on the type of data and the nature of the task, this could involve costs of specialized equipment and labor, or contracting an external company. The cost of gathering training data must be considered in the cost of implementing an AI tool. The costs associated with storing data both locally and on the cloud must also be factored in, especially with large volumes of SCADA data and high-resolution images that will have large file sizes.

Energy

Given the computing power involved in training and running large AI models, the energy consumption is a major factor in the cost of running AI tools. Small models may be able to run on a consumer-level device, while large models running on HPCs will consume large amounts of energy and will also require energy and water for cooling.

The Department for Energy Security & Net Zero (2023) predicts that the LCoE of OSW generation will remain relatively stable through to 2040, with estimates of £32–51 per MWh in 2040, depending on the level of capital expenditure of the wind farm, with onshore wind energy showing a slight reduction in cost to £31–40 per MWh, and large-scale solar showing a significant reduction to £26–36 per MWh in the same time frame. Meanwhile, the LCoE of combined cycle gas turbine (CCGT) generation is expected to skyrocket to £178–181 per MWh by 2040. If these predictions are borne out, OSW will become an increasingly economical option for powering AI data centers.

The Jevons Paradox occurs when an increase in efficiency leads to a rise in uptake, meaning that resource consumption stays the same or even increases. In the first industrial revolution, this occurred when more efficient machinery led to an increase in demand for coal. Chen (2025) suggests that while more energy-efficient AI tools may require less energy to run, if the uptake of these tools is faster than reduction in power demands, then overall energy use associated with AI will continue to increase.

Training

Generalized pretrained models can help to reduce the setup costs of AI models, with smaller amounts of data needed to fine-tune models to the specific task, and with less training time and energy required. Pretrained weights for the popular object detection task COCO (Lin et al., 2015) are available for a range of models in PyTorch (Paszke et al., 2019). These models can then be fine-tuned for defect detection tasks with small amounts of annotated data. The main limitation of using pretrained models is that not all models have pretrained weights available. Furthermore, pretrained weights are specific to the model architecture, meaning that models cannot be significantly modified if pretrained weights are to be used.

Foundation models such as OpenAI's GPT series (OpenAI, 2022; OpenAI et al., 2024) and Meta's LLaMa series (Touvron et al., 2023) are popular in generative AI tasks, as they are pretrained on vast amounts of data, and therefore can be applied to a broad range of tasks. As with pretrained object detectors, these models can be fine-tuned to more specialized tasks, although careful prompt engineering can reduce the need for fine-tuning without the requirement for specific training data.

Retrieval augmented generation (RAG), proposed by Lewis et al. (2020), allows LLMs to access external knowledge bases to improve the quality of their responses. When queried, the model will retrieve information from relevant sources, known as chunks, and use these to supplement the LLMs' internal knowledge base to generate a more relevant response. Instead of retraining an LLM on a new corpus, the knowledge base can be updated, saving computational cost compared to the standard approach of re-training from scratch. In the OSW industry, a RAG-based model could allow a question-answering system to retrieve and present the procedures for a certain maintenance operation from documents such as service manuals.

Recommendations

To reduce the training requirements of models, we encourage the use of transfer learning and fine-tuning where possible, using adapted pretrained models from related tasks to new domains. Transfer learning reduces both the data requirements of the model and the amount of training required, tackling two major contributors to the cost of AI tools. Alternatives to retraining such as RAG should also be considered where appropriate.

We also encourage the use of the smallest and simplest AI models required when building a tool. Although the latest and largest LLM foundation model may be appealing, if a more energy-efficient model is capable of adequate performance, then the additional cost and environmental impact should be carefully weighed against the benefits. Similarly, the computing hardware should be carefully considered to meet the needs of current and foreseeable models, with some sensible allowance for future-proofing to prevent hardware from becoming prematurely obsolete.

If the widespread adoption of AI tools in the OSW industry can contribute to a reduction in LCoE, while becoming an increasing component of the energy mix, then the costs of powering AI tools will decrease. This would likely increase adoption, creating a symbiotic cycle between AI and renewable energy. In the OSW industry, this could be further exploited by co-siting AI data centers with onshore infrastructure such as substations and powering directly from the renewable source when available.

11.2.2 O&M Costs

For businesses, the main driver of change is the hope of return on investment. Operators and O&M firms will adopt AI tools that increase their profitability over existing methods, such as by reducing costs, increasing availability, and giving an edge over competitors in the hope of winning future bids or contracts. Meanwhile, consumers will support the adoption of these tools if they help to reduce the cost of energy.

Estimates place the cost of O&M at between 19% and 33% of the total life-cycle cost of a wind farm (Abu-Rumman et al., 2017; Costa et al., 2021; Hammond and Cooperman, 2022), making potential savings from adopting AI tools especially attractive to operators. The costs of O&M include transport to the worksite, turbine downtime, insurance for hazardous operations including working offshore and at height, in addition to the costs of materials and labor. The integration of AI tools into O&M has the potential to reduce these costs by reducing the frequency and duration of operations, reducing turbine downtime, reducing the need for staff to access blades for inspections, and by preventing serious failures from occurring through early intervention.

Maximizing Availability

In addition to reducing the costs of O&M, AI methods can also increase the profitability of OSW generation by increasing availability, optimizing generation to demand, and automating the bidding process. To maximize profits, it is important that OSW farms are able to maximize availability and minimize downtime.

When performing correctly, the power output of a wind turbine can be determined by a relatively simple equation based on the wind speed, along with the cut-in speed and rated power. This means that accurate forecasting of wind speeds go hand in hand with accurate forecasting of power generation. Modern AI TSF methods are able to learn time-series features at a number of scales, including seasonal, daily, and hourly trends, making them well-suited to the prediction of weather conditions where statistical and simulation-based approaches would be unsuitable.

Tuton et al. (2023) apply a kNN regression-based tool to the prediction of wind speed for the OSW industry, upscaling wind speed forecasts at 10 m above sea level to WT hub height for forecasting of power generation, achieving encouraging results forecasting a week ahead with a relatively simple model. Figure 11.2 shows examples of the predicted wind speed in m/s and power generation in kWh vs the actual values from SCADA data, demonstrating the generally good performance of the kNN model.

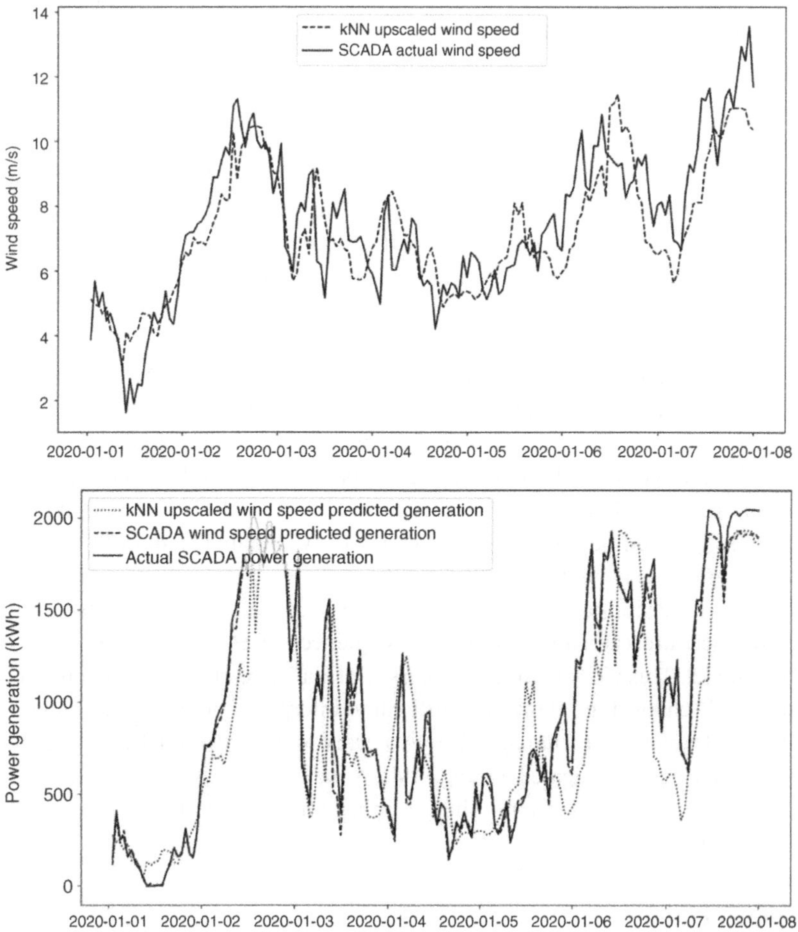

Figure 11.2 Seven-day kNN upscaled wind speed (m/s) and power generation (kWh) predictions compared to actual and predicted values from SCADA data (Tuton et al. [2023] / EDP Sciences / CC BY 4.0).

More complex ML and DL models such as neural networks, recurrent neural networks (RNNs) including long short-term memory (LSTM) and gated recurrent unit (GRU) networks have also been applied to TSF tasks for both weather and power generation forecasting, showing promising results (Lin and Liu, 2020; Kisvari et al., 2021; Wei and Chiang, 2024).

More recently, Transformer models have revolutionized the training and use of TSF models, running in less time than previous methods such as Recurrent NNs, while learning key features in the data using the multi-head attention mechanism. Promising results have been shown in applications in the OSW industry by Mo et al. (2024) and Wan et al. (2024). The adoption of AI models for forecasting at a turbine and wind farm level has the potential to improve the optimization of generation, by giving more detailed and accurate forecasts further ahead of time, and updating in real time.

Maintenance Planning

In order to plan O&M operations efficiently, it is beneficial to be able to predict faults before they arise, minimizing costs and downtime. Reliable TSF is key to the field of Predictive Maintenance, where the future state of an asset is predicted based on previous and current states, allowing maintenance work to be scheduled before a failure occurs.

As part of the shift toward Smart O&M in the OSW, fault prediction tools have already been implemented in some applications. ORE Catapult (2024) reports that AI-based tools are able to predict faults on WTs up to 30 days before they occur, which allows for improved O&M planning, potentially saving £4000–15 000 per day of turbine downtime. Chrétien et al. (2024) present a method for blade damage prediction using SCADA data and aeroelastic modeling, and Chandrasekhar et al. (2021) present a methodology for the detection of WTB damage up to six months ahead of manual inspection.

Taking a different approach to predicting events from SCADA data, Walker et al. (2022) predict future repair actions from sequences of SCADA alarms using a BiLSTM model. This work highlights the issue of alarm overload, where nuisance and repeated "chattering" alarms are difficult to interpret, and makes recommendations for a framework to manage SCADA alarms. Recent research by the authors of this chapter has investigated adding PIML principles to existing TSF methods, infusing known physical laws into their training and predictions to predict crack growth in WTBs under loading.

Recommendations

Improvements in AI methods for time-series forecasting of weather conditions and power generation, and for the prediction of faults, have shown great potential in a number of publications, but their application in the industry has been limited so far. We consider that the economic arguments for adoption are strong, with potential savings in O&M costs being attractive to operators, and O&M firms wishing to remain competitive with rivals, and that the performance of leading models is highly encouraging compared to existing methods. However, as the underlying AI technology is relatively untested in critical applications, we understand that uptake will be gradual at first, increasing over time as technologies mature.

11.2.3 Environmental Factors

The combination of the proposed benefits of AI for the OSW industry and directions of research in renewable energy and reducing the environmental impacts of AI show significant potential for mutual benefits. The environmental impacts of AI tools are difficult to quantify, and include direct and indirect impacts.

CO_2 Equivalent

Due to the power requirements for training and running AI models, the carbon footprint of AI has long been a contentious issue. Before the widespread adoption of LLMs, the training stage was considered to be the primary contributor to the environmental impact of AI models. However, the computational demands of large numbers of users running generative AI now mean that the inference stage often consumes more resources.

Yu et al. (2024) estimate that the CO_2 equivalent (CO_2e) emissions from leading AI systems could soon reach 102.6 Mt per year, greater than that of many countries. Google's Gemini Ultra is reported as generating 37 620 t CO_2e in a single training run, while the emissions from training GPT-4 have increased by a factor of 12 from GPT-3.5 to 21,660 t. For comparison, per capita CO_2e emission worldwide is estimated at 4.7 t CO_2e per annum. Clearly, the environmental impacts of AI systems cannot be ignored.

The estimates provided by Yu et al. (2024) are based on coal as the primary power source this trend is illustrated in Figure 11.3. The National Renewable Energy Laboratory (2013) estimates that wind power generates around 10 g CO2e per kWh, while coal generates around 1000 g. Although the full adoption of renewable energy sources is highly optimistic, if a significant portion of AI systems can be powered by renewables then a considerable impact can be made in terms of carbon footprint during training and runtime.

Water Cooling

In large data centers, servers are cooled using cold water to prevent overheating. Li et al. (2025) place the water requirement for running a data center at around 2 L/kWh of electricity used. Estimates place global water demand for AI as high as 6.6 billion cubic meters by 2027, almost half of the water usage of the United Kingdom. Manufacturing of electronics is also water intensive, adding to

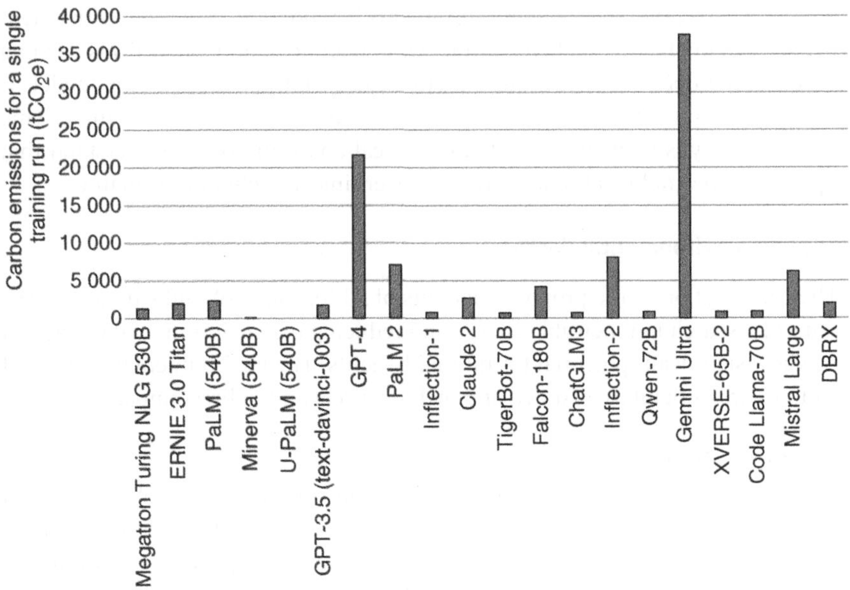

Figure 11.3 Carbon emissions (CO_2 equivalent) for a single training run of the top 20 highest emitting AI systems (Yu et al. [2024] / Springer Nature).

the total life-cycle demands of AI tools, which must be considered when setting up and refitting data centers.

Hardware and Software Life Cycles

The rapid pace of AI development means that both software and hardware become obsolete quickly. For software, this means that increasingly complex models must be trained, on larger amounts of training data than their predecessors, increasing the environmental impact of training with each generation.

For hardware, this means that equipment, such as GPUs, is replaced frequently to run these larger models at increasing speed. This creates large amounts of e-waste, which contains considerable amounts of toxic heavy metals. The life-cycle environmental impacts of computing equipment include mining and processing raw materials, manufacturing, shipping, and recycling, in addition to the power and cooling resources when in use.

Recommendations

Many of the recommendations made around reducing the setup and running costs of AI tools also have environmental benefits, adding additional incentives to carefully considering how AI tools are selected and implemented. By minimizing the power consumption, the running costs, carbon footprint, and water use of AI tools can be reduced. Luccioni et al. (2025) publish a leaderboard of energy scores for popular AI frameworks in a range of tasks, including generative AI and vision tasks, providing GPU energy usage and a score from 1 to 5 stars. We recommend that a leader board such as this is consulted when selecting models so that energy use can be considered along with performance and other factors.

Careful consideration of the equipment life cycle will reduce impacts associated with e-waste by prolonging the lifetime of computing equipment. However, this should be balanced with the actual requirements of current models, as running overpowered equipment will consume more energy than necessary. Looking further ahead, if future AI data centers for wind farm monitoring can be directly powered using renewable energy from the source, then their carbon footprint from energy use can be reduced to near zero, except for in periods where generation is not possible due to low winds or extreme weather.

11.2.4 Deploying AI in the Field

Due to the complexity of AI models, a considerable amount of computing power is required to run them at speed. Many large AI models are trained and run on GPU clusters, or remotely via cloud computing. This presents an issue for offshore work, as portable devices have far lower computing power than the HPCs that models are trained on, and internet connection on wind farms relies on mobile communications, if available at all. The field of Edge AI explores the deployment of lightweight AI models on platforms with less computing power than conventional GPUs and HPC clusters.

The increasing processing power of mobile devices has enabled deployment of AI on mobile devices such as laptops, tablets, and smartphones, which can be used in the field without requiring an internet connection to a server. Embedded computing boards equipped with GPUs are also becoming increasingly accessible, with the NVIDIA Jetson series offering up to 275 TOPS on a postcard-sized module (NVIDIA, n.d.). These miniaturized boards are intended for integration into autonomous devices, and have been able to run object detection tasks in real time (Hakani and Rawat, 2024). Popular AI frameworks such as TensorFlow (Abadi et al., 2015) have also been deployed on Raspberry Pi devices, which have been demonstrated on miniature self-driving vehicles by Tung (2018).

Recently, work in defect detection on WTBs has focused on running ML models on edge devices such as laptops and tablets for use in the field and real-time processing on video feeds in the field. Lightweight versions of the popular object detection network YOLO have been run on NVIDIA Jetson edge computing platforms at 23–83 FPS (Hu et al., 2023; Wu et al., 2024) while maintaining high accuracy, and a YOLO model has been integrated into a mobile app to be run on a consumer-level device for wind turbine inspection (Zou et al., 2025).

Recommendations

In the near future, the combination of increased mobile computing power and efficient networks will continue to increase the AI capabilities of mobile devices, while reducing barriers to the adoption of AI tools for offshore operations. As with reducing costs and environmental impacts of running AI tools, we recommend that models and hardware are selected based on the actual requirements of the task, instead of running more complex models on overpowered GPUs for a slight improvement in processing speed.

11.2.5 Dependability/Trustworthiness

One of the most fundamental concerns around current AI models is their dependability in critical applications. Most ML and DL models operate as "black boxes," where their internal processes, known as hidden states, are unseen by the user and are generally not interpretable to humans, who only see the inputs and outputs. When AI models are trained well, they will show high performance on a task based on certain metrics, and in many applications this is sufficient for them to be trusted and applied with minimal oversight. However, in applications where mistakes could be costly or even dangerous, more oversight is required.

In the OSW industry, the biggest concerns around the dependability of AI are missed faults and damage leading to failures, inaccurate forecasting leading to insufficient power generation, and the safety of workers offshore.

Trustworthy and Explainable AI

Trustworthy and Explainable AI are fields of AI research that aim to tackle these concerns. Trustworthy AI aims to increase trust in AI tools in critical applications by making models fairer, more reliable, and more interpretable to users. Explainable AI, or XAI, aims to give users insight into the reasoning of AI models, lightening the black boxes of ML and DL models. When a model makes an incorrect prediction or decision, XAI can aid in determining why the model has learnt this behavior, and allow this to be rectified.

A popular form of XAI is the use of visualizations in computer vision (CV) tasks, where pixels and regions that contribute strongly to a classification are highlighted, often in the form of a heatmap. Highlighting key features can shed light on features in an image that may be misinterpreted. In inspection tasks, these visualizations can be especially powerful, as they can highlight potential faults for human oversight.

One such method is gradient-weighted class activation mapping (Grad-CAM), proposed by Selvaraju et al. (2019). Grad-CAM can be applied to a wide range of CNN-based CV models, including image classification and captioning tasks. An example output from Grad-CAM applied to a WTB damage detection task by

Iyer et al. (2022) is shown in Figure 11.4, showing areas in an image that contribute strongly to predicted areas of damage (HIT).

In attention-based models such as Transformers, the most relevant features will be focused on, and identifying these features can shed light on the underlying logic behind predictions. More recently, flexible XAI tools have been developed that can be applied to black-box models, including SHapley Additive exPlanation (SHAP) for prediction tasks (Lundberg and Lee, 2017), and Local Interpretable Model-Agnostic Explanations (LIME) for classification tasks with two or more classes (Ribeiro et al., 2016). Liao et al. (2024) review the application of XAI in wind speed forecasting, including LIME and SHAP, and show encouraging results on a case study on real-wind farm data.

Recommendations

We consider that trustworthy AI and XAI methods offer the potential to tackle one of the key issues in the adoption of AI tools in industrial tasks. By providing insight into AI models, users will be better able to evaluate the quality of the outputs, and trust in high-performing models will naturally increase. Liao et al. (2024) point out that many XAI methods, including SHAP, are computationally demanding, especially when applied to large and complex models such as CNNs. Therefore, this is encouraged as an area of research in the near future to ensure that XAI methods can be applied where required.

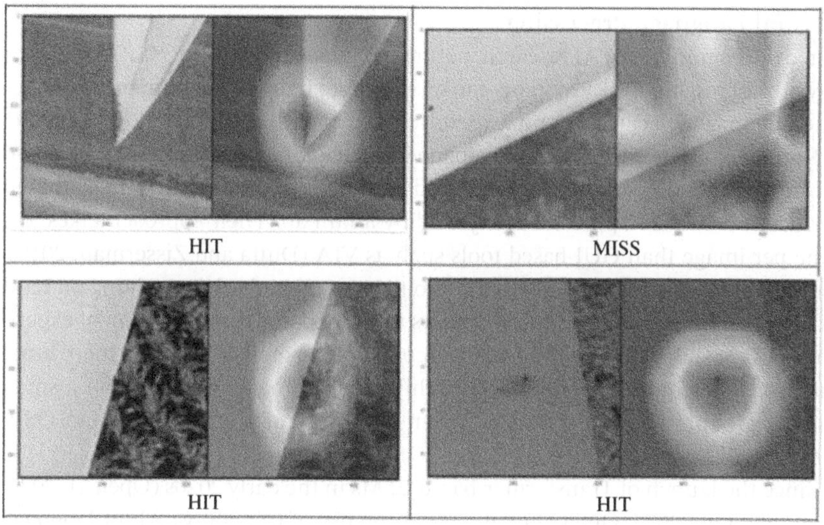

Figure 11.4 Example Grad-CAM explanations on a wind turbine blade damage detection task (Iyer et al. (2022) / arXiv / CC BY-NC-ND 4.0).

11.2.6 Human–AI Interaction

In O&M tasks in the OSW industry, we consider that it is unlikely that a significant amount of maintenance decisions and operations will be fully automated, instead being used as a decision support tool to supplement the expertise and experience of human staff. Therefore, the interaction between humans and AIs is an important area of research to ensure that the use of AI is optimized.

Human-in-the-loop ML

AI tools are well-suited to laborious and repetitive tasks, including monitoring of continuous data and processing large amounts of data. While humans may interpret data differently or make mistakes, especially when bored or fatigued, AI tools are tireless and consistent. However, AI tools are limited in their capacity to generalize, which can cause issues when previously unseen features arise in the data, an area in which humans excel. Human-in-the-loop (HITL) ML is a field of AI research that focuses on AI tools which receive a level of human input, during training or at runtime. This can include interactive learning, where control over training is shared between the AI and a human who provides input, and forms of reinforcement learning where human-approved outputs are rewarded and training is updated over time. While implicit human oversight exists in the form of approving or ignoring predictions from an AI tool, we consider that this oversight should be used to improve models.

Natural Language Processing

Since the beginnings of AI research, natural language processing (NLP) has been a key area of research. NLP systems can enable more naturalistic human–computer interaction, which can allow nonexperts to interact with systems with greater ease.

Rothon et al. (2024) present a dialogue interface for gathering human input on an interactive ML task, where predicted instances of damage in UAV-gathered images of a WT are approved or rejected by a human. Their approach takes less time per image than GUI-based tools such as VIA (Dutta and Zisserman, 2019), and requires fewer data points to train to the same standard, by training on hard examples. This approach reduces barriers to the expertise and intuition of experienced O&M staff when training AI tools compared to methods that are more time-intensive, or require expertise in AI. While this approach was based on a small sample group, it demonstrates the potential of interactive learning in an OSW O&M task.

Since the launch of Transformer-based LLMs in the early 2020s (OpenAI, 2022; Touvron et al., 2023; OpenAI et al., 2024), the use of Generative AI for language tasks has become part of daily life. The use of LLMs for decision support in O&M has been proposed, with a number of publications showing success in generating clear explanations of engineering procedures. However, LLMs have been found to have issues with the quality and reliability of generated content, outputting incorrect, irrelevant, and potentially unsafe responses, known as hallucinations (Levy et al., 2022; Huang et al., 2024; Dong et al., 2024).

In the OSW industry, Walker et al. (2023, 2024a,b) have investigated detecting and filtering hallucinations and unsafe responses in LLM outputs when questioned on O&M procedures. Walker et al. (2024a) highlight the need for domain-specific data from industry to train a successful model. Methods including RAG (Lewis et al., 2020; Lyu et al., 2024) and Knowledge Graphs (KGs) (Jaffri, 2024; Zhu et al., 2025) have also been proposed to improve LLM-generated output by basing responses on approved sources.

Recommendations

We believe that through improvements in NLP, interaction between humans and AI tools can be made more natural, especially in cases where AI nonexperts are interacting with complex systems. KG- and RAG-based dialogue systems offer the ability to tackle the current issues with LLMs around hallucinations and unsafe responses, while imbuing domain-specific knowledge into models without additional training, reducing other concerns about the cost and resource use during setup. HITL ML and RL approaches also show promise in adding human knowledge into model training, with the strengths and weaknesses of humans and AI tools complementing each other.

In cases where improving the outputs of models fails, then filtering for unsafe prompts and outputs should be as a last resort. We encourage researchers to continue to present both novel architectures and applied case studies to demonstrate the feasibility of these approaches to the industry, while also carefully and critically considering the use cases and potential pitfalls of use of AI-generated content in safety-critical applications. As legislation catches up with the pace of AI development, this will also affect how AI tools can be used in potentially hazardous situations.

11.2.7 Cybersecurity

As OSW energy becomes an increasing component of the energy mix of many countries, the vulnerability of heavily computerized systems to cyberattacks is a concern. The applications of AI in improving the resistance of RE systems to cyberattacks have been explored by Barreto et al. (2020), Ghiasi et al. (2023), and Rekeraho et al. (2023).

Hoppa (2023) discusses cybersecurity threats to wind farms, highlighting vulnerabilities, including connections to public internet, wireless connections, and equipment using legacy software, which no longer receives security updates. Risks to OSW infrastructure include directory traversal, cross-site scripting, information disclosure, malware injection, and denial of service (van den Brink, 2022). Staggs et al. (2017) identify points of attack as physical access at wind turbines or substations, cyber access via vendor networks, technician equipment, and compromised supply chains. Cyberattacks on WTs and infrastructure can cause temporary blackouts, and in extreme cases can be used to sabotage equipment, causing longer-term issues (Staggs et al., 2017; van den Brink, 2022).

As increasing amounts of internet-connected systems are present in WTs and modern smart grids, more attention is being paid to Internet of Things (IoT) cybersecurity, as each device represents a potential weakness to attack. Rekeraho et al. (2023) consider the vulnerabilities of IoTs in RE as including insecure communication protocols, poor encryption methods, lack of access control, lack of parameter sanitization, and inappropriate use of authentication alongside encryption. van den Brink (2022) found at the time of writing that some WTs are still controlled by running on Windows 2000, for which support ended in 2010. As WTs have a lifetime of 20+ years it is highly likely that their systems will need to be updated multiple times to remain secure.

Recommendations

Cybersecurity measures have improved over time as the OSW industry has matured, but existing infrastructure is still a point of weakness, and threats are ever-changing. Staggs et al. (2017) and Hoppa (2023) recommend that operators engage with cybersecurity experts to strengthen their security measures, developing strong policies and procedures to mitigate the complex threats. We also consider that the use of wireless IoT devices must be justified against the security risks.

11.2.8 Data Availability

A key issue surrounding AI research in the OSW industry is the lack of publicly available data and the lack of collaboration between operators, WT manufacturers, and academia in creating datasets for research. Multiple sources note data sharing as necessary for achieving goals for renewable energy generation, and noting mutual benefits (Li et al., 2023). Currently, operators, OEMs, and inspection companies rarely publish data, including SCADA data and images from inspections, due to commercial sensitivity and reluctance to publicize the failure modes of their equipment. While this maintains profitability, it also means that little data makes it into the public domain for research purposes, and what does is often outdated. This has been noted as a key limitation when building state-of-the-art dialogue systems for highly specialized applications by Walker et al. (2024a,b), both when using classical training methods, and when using KG- and RAG-based approaches.

Leading object detection models are trained on large datasets such as ImageNet (Russakovsky et al., 2014) and COCO (Lin et al., 2015), which contain millions of examples across a range of object categories. Datasets with comparable numbers of annotated examples are not widely available for WT inspection tasks. Most publicly available datasets contain fewer than 3000 examples of damaged WTBs (Shihavuddin and Chen, 2018; Nikolov et al., 2020; Altice et al., 2024), and papers reporting state-of-the-art performance on OSW tasks such as Iyer et al. (2022) often do not publish their data.

Approaches including few-shot learning have shown success in reducing the impact of limited data availability in a range of applications, including in the OSW industry. Yu et al. (2022) propose DMNet, a framework for wind turbine surface defect detection based around FSL, which has beaten conventional CNNs in testing. DMNet uses meta-learning, which increases the ability of the model to adapt to new tasks. Data augmentation methods are capable of expanding datasets by applying transformations to the existing data to generate more variety. Patel et al. (2021) apply basic data augmentations including shifts, rotations, flips, and zooms to expand the size of a dataset of images of a section of WTB containing damage. While these approaches are useful tools where data does not currently exist, there is a wealth of data from operational wind turbines, it is simply not entering the research landscape.

Recommendations

In order to undertake cutting-edge research which pushes applications further, AI researchers need up-to-date datasets from industry to train and test their models. Therefore, we strongly encourage operators and industry bodies to facilitate the sharing of data where possible. We recognize that data from industry sources will come with concerns around security and commercial sensitivity, which may necessitate anonymization and final say over publications of results from sensitive data. We also encourage researchers to consider publishing data generated as part of their research as a valuable contribution, alongside results and code. We consider that FSL is a valid solution where data gathering is prohibitively expensive or impractical, but that it should not be used as a sticking plaster solution.

11.2.9 Collaboration Between Academia and Industry

It is clear that AI researchers wish to propose and develop innovative solutions, with potential for application. In academia, this is often seen as more "pure" or fundamental research, i.e. research to further knowledge. Meanwhile, industry-based Research and Innovation is seen as geared more toward development of tools and solutions with concrete applications and commercial potential. In reality, both academic and industry-based research are driven by a combination of factors.

Recommendations

To develop tools that fit the needs of the OSW industry, communication on research gaps and existing tools is required. The enthusiasm of academics should be harnessed by the OSW industry where possible, fostering close collaboration with academia to access both cutting-edge technologies, and the talent and expertise of researchers throughout their careers.

11.3 Communication and Opinion

In the 2020s, significant advances have been made in a range of AI fields, which has generated excitement around the future of AI, and led to rapid adoption of AI tools such as LLMs. However, it is important that new technologies are not simply applied as they are new and in vogue, but because they offer the best fit for the task at hand, considering performance and cost. We consider how the wider opinion on AI may impact its adoption.

11.3.1 AI Winters

A consideration when predicting the growth of emerging technologies is the existence of trends, where optimism for a technology is buoyed by a period of progress. If care is not taken to fully justify the application of the technology, then trends can lead to certain solutions being used more because they are currently in vogue, and not necessarily because they represent the best fit. Trends will run for a finite amount of time, and often result in a bubble. This has notably occurred before in the field of AI research, with major AI "winters" between 1974–1980 and 1987–2000 where research stagnated after periods of promise gave way to disappointment and disillusionment.

In recent years, AI research has progressed at an increasing rate due to factors including GPU acceleration and the proposal of the Transformer architecture. With the increasingly rapid pace of progress around AI and ML/DL, it is difficult to predict if and when development will stall, especially without the benefit of hindsight. Molander (2023) argues that while generative AI is growing, classical AI methods may be entering another AI winter, noting lack of progress in DL models in recent years. If a classical AI winter occurs, it is likely that research into the analytical and predictive models that are dominant in the OSW industry will slow.

11.3.2 Search Trends

An indicator of a coming slump is growing cynicism around the benefits of an emerging technology. Virtually since the introduction of Generative AI, concerns around plagiarism have been abundant. The term "AI Slop," first recorded in

2020, became popular in 2024 to describe low-quality generated content including images and text, comparable to "Spam" for unwelcome email content, popularized by Willison (2024).

Analysis of internet search trends from Google Trends (n.d.) since the launch of ChatGPT in November 2022 shows that searches for "AI" have increased by around a factor of 7. Meanwhile, searches for negative terms around AI including for "AI Plagiarism," "AI Fake," "AI Propaganda," "AI Slop," and "AI Bad" have increased significantly since late 2022, shown in Figure 11.5. Searches for "AI Plagiarism" increased by a factor of around 30, peaking in late 2024. Similarly, searches for "AI Propaganda" have grown since 2022, indicating increasing awareness of political use of AI. The increasing popularity of terms such as this signals public disillusionment with generative AI, although how deeply this affects the adoption of AI tools remains to be seen.

11.3.3 Gartner Hype Cycle

Dörterler et al. (2024) recommend that in addition to NNs and ML methods, techniques including fuzzy logic, data mining, and statistical methods should be considered to suit specific circumstances. These methods make up 27% of citations identified, while artificial neural networks (ANNs) and ML methods constitute 59%. One model through which the growth of emerging technologies can be considered is the Gartner Hype Cycle (Jaffri, 2024), which consists of five phases.

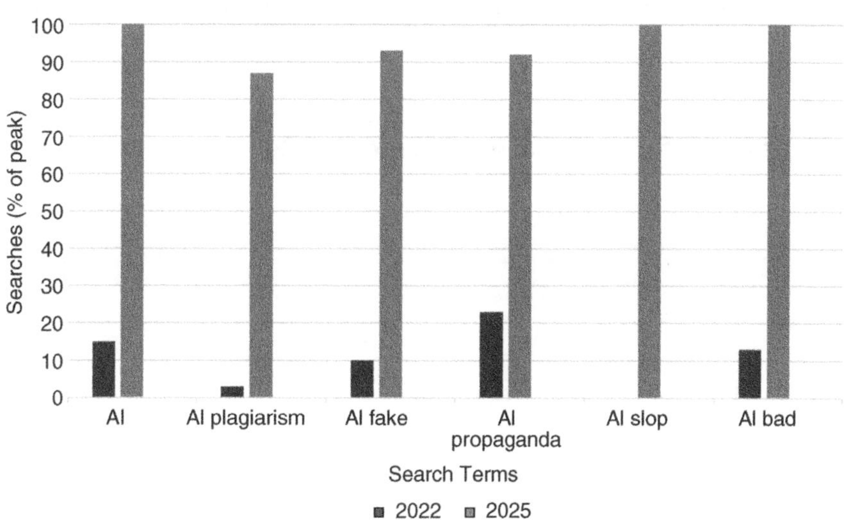

Figure 11.5 Google searches for negative sentiments on AI November 2022 to April 2025.

1. **Technology Trigger:** The emergence of a new technology begins the cycle.
2. **Peak of Inflated Expectations:** Early successes lead to high expectations.
3. **Trough of Disillusionment:** Interest fades as expectations are not met.
4. **Slope of Enlightenment:** Technology begins to mature as subsequent generations emerge and the benefits and drawbacks become more understood.
5. **Plateau of Productivity:** Mainstream adoption occurs, and criteria for assessment become clear. If the technology has a broad market then it is likely to continue to grow over time.

In the Gartner Hype Cycle for Artificial Intelligence 2024, published by Jaffri (2024) and shown in Figure 11.6, Generative AI is placed past the "Peak of Inflated Expectations," moving into the "Trough of Disillusionment," and predicted at 2–5 years until the "Plateau of Productivity" is reached. While Generative AI is being adopted at a rapid rate due to its broad capabilities and ease of access, signs of dissatisfaction are beginning to show in public opinion.

Computer Vision is placed close to the "Plateau of Productivity," and is predicted to plateau within two years. Considering the near-human level performance shown in terms of accuracy and speed, this seems reasonable, although the exact timeline and metrics could be debated. Similarly, Edge AI is considered close to plateau, although due to enter the "Trough of Disillusionment" phase.

The Gartner Hype Cycle has been the subject of debate, with critics including Molander (2023) pointing out that many technologies do not fit in the prescribed cycle. While the Gartner Hype Cycle may not be a perfect model, it can be an interesting way to consider how AI technology may progress over the next decade.

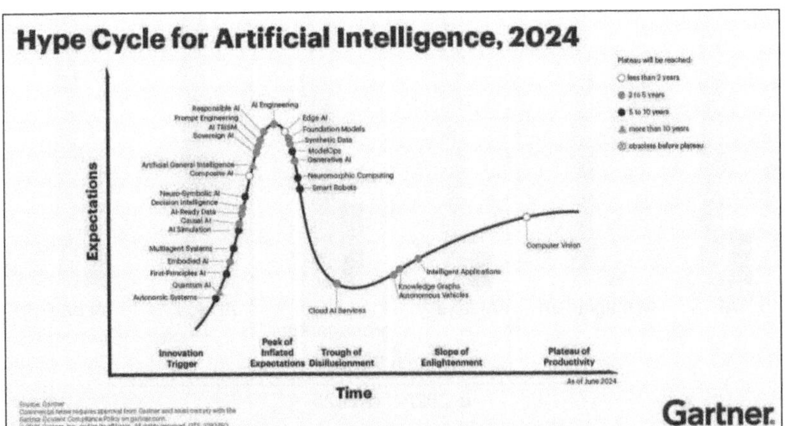

Figure 11.6 Gartner Hype Cycle for AI, 2024 edition (Jaffri, 2024).

11.3.4 Conflicting Findings and Definitions

Due to the rapid pace of progress in AI research in the past decade, and the growing number of publications on increasingly complex and diverse methods available, conflicting views and results are common. Disagreements often arise around the benefits of novel methods over older methods and around terminologies.

Digital twins (DTs) are a growing area of research, where a physical system is represented by a computerized version, with varying levels of connection. Forecasting DTs have the capacity to predict future behavior of an asset through combinations of ML and simulation. In the OSW industry, this has been applied to a range of tasks including power generation forecasting and failure prediction.

One of the key factors impeding the widespread adoption of DTs is the conflicting definitions found in the literature. This has resulted in a range of classifications, where specialized DTs are grouped by varying characteristics under the umbrella of DTs. Efforts have been made by AMRC (2020) and Tuton et al. (2024) to better define what characterizes a DT, with the use of industry-specific definitions being proposed in order to decrease confusion. Until a consensus on the nomenclature of DTs can be reached, it is likely that the field will suffer from approaches being misclassified, misapplied, or simply ignored.

11.3.5 Overstated Benefits of Novel Methods

A factor that risks contributing to a downturn in AI growth is the overstatement of benefits of AI over existing methods, which may be simpler, leading to disappointment and cynicism around future claims. One example of this is presented by Zeng et al. (2022), which aims to encourage debate around the benefits of Transformer models for long-term TSF tasks, demonstrating superior performance with a set of "embarrassingly simple" one-layer linear models. The findings of Zeng et al. (2022) were later challenged by Simhayev et al. (2023), who report superior performance by a simple transformer model.

11.3.6 Recommendations

In order to minimize the confusion of nonexperts who are simply seeking to adopt the most effective and appropriate tool for their needs, we call on AI researchers to be open and honest about the limitations of their methods, as well as publicizing the benefits. As part of this, we believe that effective communication is necessary with nonexperts in plain language, avoiding jargon while succinctly explaining how the technology works.

In order to prevent a downturn in AI research, we consider open and honest communication around the strengths and failings of AI at all levels. At present, many users of AI have little understanding of the core technology that powers the tools that they use, and therefore are not aware of the limitations and failure modes of these tools. By communicating better with nonexperts, using less technical language and jargon, AI researchers can better engage the public. Before previous AI winters, overinflated expectations from overly optimistic predictions led to disappointment and disillusionment with AI, and it falls to the research community to prevent this from repeating. While our survey of search trends is far from exhaustive, the results illustrate the growth in public sentiment against AI.

11.4 Conclusion

In this chapter, we have discussed how multiple issues around the adoption of AI tools in OSW O&M may be tackled, considering key stakeholders. We find that while AI tools have begun to be adopted in the OSW industry, concerns around the economics, performance, and security must be addressed before these tools can reach their full potential. We also consider that nontechnical concerns around the wider use of AI and relationships between industry and academia are impacting the pace of development and uptake of AI tools.

The key recommendations that we make to the industry (operators, O&M contractors, industry researchers) and academia are around collaboration and openness. Only by fostering close communication and collaboration can industry-ready AI tools be developed and implemented to meet sustainability targets. We also urge all stakeholders to weigh the costs and benefits of AI methods for the use cases, considering recent ML and DL methods alongside simpler AI methods, and against non-AI methods. Communication between experts from the industry and academia will allow closer collaboration to resolve technical concerns, while better communication around AI to the public will improve public trust in AI.

Bibliography

Martín Abadi, Ashish Agarwal, Paul Barham, Eugene Brevdo, Zhifeng Chen, Craig Citro, Greg S Corrado, Andy Davis, Jeffrey Dean, Matthieu Devin, et al. TensorFlow: Large-scale machine learning on heterogeneous systems. *arXiv preprint arXiv:1603.04467*, 2015. https://www.tensorflow.org/.

Ala' K Abu-Rumman, Iyad Muslih, and Mahmoud. A Barghash. Life cycle costing of wind generation system. *Journal of Applied Research on Industrial Engineering*, 4(3):185–191, 2017. https://doi.org/10.22105/jarie.2017.54726.

Bridger Altice, Edwin Nazario, Mason Davis, Mohammad Shekaramiz, Todd K Moon, and Mohammad AS Masoum. Anomaly detection on small wind turbine blades using deep learning algorithms. *Energies*, 17(5):982, 2024. https://doi.org/10.3390/en17050982.

AMRC. Untangling the requirements of a digital twin, 2020. https://www.amrc.co.uk/files/document/406/1605271035_1604658922_AMRC_Digital_Twin_AW.pdf. (accessed 31 March 2025).

Carlos Barreto, Taha Eghtesad, Scott Eisele, Aron Laszka, Abhishek Dubey, and Xenofon Koutsoukos. Cyber-attacks and mitigation in blockchain based transactive energy systems. In *3rd IEEE International Conference on Industrial Cyber-Physical Systems*. IEEE, 2020. https://doi.org/10.1109/ICPS48405.2020.9274708.

BVGAssociates. *Guide to an Offshore Wind Farm*. BVG Associates, 2019. https://guidetoanoffshorewindfarm.com/.

Kartik Chandrasekhar, Nevena Stevanovic, Elizabeth J Cross, Nikolaos Dervilis, and Keith Worden. Damage detection in operational wind turbine blades using a new approach based on machine learning. *Renewable Energy*, 168:1249–1264, 2021. https://doi.org/10.1016/j.renene.2020.12.119.

Sophia Chen. How much energy will AI really consume? The good, the bad and the unknown. 2025. https://www.nature.com/articles/d41586-025-00616-z. (accessed 17 April 2025).

Antoine Chrétien, Antoine Tahan, and Francis Pelletier. Wind turbine blade damage evaluation under multiple operating conditions and based on 10-min SCADA data. *Energies*, 17(5):1202, 2024. https://doi.org/10.3390/en17051202.

Ángel M Costa, José A Orosa, Diego Vergara, and Pablo Fernández-Arias. New tendencies in wind energy operation and maintenance. *Applied Sciences*, 11(4):1386, 2021. https://doi.org/10.3390/app11041386.

Department for Energy Security & Net Zero. Electricity generation costs 2023, 2023. http://www.gov.uk/government/publications/electricity-generation-costs-2023. (accessed 29 April 2025).

Zhichen Dong, Zhanhui Zhou, Chao Yang, Jing Shao, and Yu Qiao. Attacks, defenses and evaluations for LLM conversation safety: A survey. In *Proceedings of the 2024 Conference of the North American Chapter of the Association for Computational Linguistics: Human Language Technologies (Volume 1: Long Papers)*, pp. 6734–6747. Association for Computational Linguistics, 2024.

Abhishek Dutta and Andrew Zisserman. The via annotation software for images, audio and video. In *Proceedings of the 27th ACM International Conference on Multimedia*, MM '19. ACM, 2019. https://doi.org/10.1145/3343031.3350535.

Safa Dörterler, Seyfullah Arslan, and Durmuş Özdemir. Unlocking the potential: A review of artificial intelligence applications in wind energy. *Expert Systems*, 41(12):e13716, 2024. https://doi.org/10.1111/exsy.13716.

Mohammad Ghiasi, Taher Niknam, Zhanle Wang, Mehran Mehrandezh, Moslem Dehghani, and Noradin Ghadimi. A comprehensive review of cyber-attacks and defense mechanisms for improving security in smart grid energy systems: Past, present and future. *Electric Power Systems Research*, 215:108975, 2023. https://doi.org/10.1016/j.epsr.2022.108975.

Google Trends. Google trends, n.d. https://trends.google.com/trends/. (accessed 13 March 2025).

Raj Hakani and Abhishek Rawat. Edge computing-driven real-time drone detection using YOLOv9 and NVIDIA Jetson nano. *Drones*, 8(11):680, 2024. https://doi.org/10.3390/drones8110680.

Ahmad Hamdan, Kenneth Ibekwe, Valentine Ilojianya, Sedat Sonko, and Emmanuel Etukudoh. AI in renewable energy: A review of predictive maintenance and energy optimization. *International Journal of Science and Research Archive*, 11:718–729, 2024. https://doi.org/10.30574/ijsra.2024.11.1.0112.

Rob Hammond and Aubryn Cooperman. Windfarm operations and maintenance cost-benefit analysis tool (WOMBAT). In *NREL is a National Laboratory of the U.S. Department of Energy Office of Energy Efficiency & Renewable Energy Operated by the Alliance for Sustainable Energy, LLC,* Technical Report. National Renewable Energy Laboratory, 2022.

Mary Ann Hoppa. Understanding cybersecurity risks in offshore wind farms. In *Proceedings of Association of Computer Science Departments at Minority Institutions Conference (ADMI'23)*. ACM, Virginia Beach, VA, USA, 2023. https://www.admiusa.org/admi2024/ADMIProceedings/FP3.pdf.

Yunzhang Hu, Liwen Wang, Tianhao Kou, and Minshu Zhang. YOLO-tiny-attention: An improved algorithm for fault detection of wind turbine blade. In *2023 8th International Conference on Intelligent Computing and Signal Processing (ICSP)*, pp. 1228–1232, 2023. https://doi.org/10.1109/ICSP58490.2023.10248560.

Lei Huang, Weijiang Yu, Weitao Ma, Weihong Zhong, Zhangyin Feng, Haotian Wang, Qianglong Chen, Weihua Peng, Xiaocheng Feng, Bing Qin, et al. A survey on hallucination in large language models: Principles, taxonomy, challenges, and open questions. *ACM Transactions on Information Systems*, 43(2):1–55, 2024.

Akshay Iyer, Linh Nguyen, and Shweta Khushu. Learning to identify cracks on wind turbine blade surfaces using drone-based inspection images. *arXiv preprint arXiv:2207.11186*, 2022. https://arxiv.org/abs/2207.11186.

Afraz Jaffri. Explore beyond GenAI on the 2024 hype cycle for artificial intelligence, 2024. https://www.gartner.com/en/articles/hype-cycle-for-artificial-intelligence. (accessed 07 April 2025).

Adam Kisvari, Zi Lin, and Xiaolei Liu. Wind power forecasting – a data-driven method along with gated recurrent neural network. *Renewable Energy*, 163:1895–1909, 2021. https://doi.org/10.1016/j.renene.2020.10.119.

Sharon Levy, Emily Allaway, Melanie Subbiah, Lydia Chilton, Desmond Patton, Kathleen McKeown, and William Yang Wang. Safetext: A benchmark for exploring physical safety in language models. *arXiv preprint arXiv:2210.10045*, 2022.

Patrick Lewis, Ethan Perez, Aleksandra Piktus, Fabio Petroni, Vladimir Karpukhin, Naman Goyal, Heinrich Küttler, Mike Lewis, Wen-tau Yih, Tim Rocktäschel, et al. Retrieval-augmented generation for knowledge-intensive NLP tasks. In *Advances in Neural Information Processing Systems*. Curran Associates Inc., Red Hook, NY, USA, volume 33, pp. 9459–9474, 2020.

Pengfei Li, Jianyi Yang, Mohammad A Islam, and Shaolei Ren. Making AI less "thirsty": Uncovering and addressing the secret water footprint of AI models, 2025. https://arxiv.org/abs/2304.03271.

Yan-Fu Li, Huan Wang, and Muxia Sun. ChatGPT-like large-scale foundation models for prognostics and health management: A survey and roadmaps. *Reliability Engineering & System Safety*, 243:109850, 2023.

Wenlong Liao, Jiannong Fang, Lin Ye, Birgitte Bak-Jensen, Zhe Yang, and Fernando Porte-Agel. Can we trust explainable artificial intelligence in wind power forecasting? *Applied Energy*, 376:124273, 2024. https://doi.org/10.1016/j.apenergy.2024.124273.

Tsung-Yi Lin, Michael Maire, Serge Belongie, Lubomir Bourdev, Ross Girshick, James Hays, Pietro Perona, Deva Ramanan, C Lawrence Zitnick, and Piotr Dollár. Microsoft coco: Common objects in context. In *European Conference on Computer Vision,* pp. 740–755. Springer International Publishing, 2015.

Zi Lin and Xiaolei Liu. Wind power forecasting of an offshore wind turbine based on high-frequency SCADA data and deep learning neural network. *Energy*, 201:117693, 2020. https://doi.org/10.1016/j.energy.2020.117693.

Sasha Luccioni, Boris Gamazaychikov, Emma Strubell, Sara Hooker, Yacine Jernite, Carole-Jean Wu, and Margaret Mitchell. AI energy score leaderboard, 2025. https://huggingface.co/spaces/AIEnergyScore/Leaderboard.

Scott Lundberg and Su-In Lee. A unified approach to interpreting model predictions, 2017. https://arxiv.org/abs/1705.07874.

Yuanjie Lyu, Zhiyu Li, Simin Niu, Feiyu Xiong, Bo Tang, Wenjin Wang, Hao Wu, Huanyong Liu, Tong Xu, and Enhong Chen. Crud-rag: A comprehensive chinese benchmark for retrieval-augmented generation of large language models, 2024. https://arxiv.org/abs/2401.17043.

Site Mo, Haoxin Wang, Bixiong Li, Zhe Xue, Songhai Fan, and Xianggen Liu. Powerformer: A temporal-based transformer model for wind power forecasting. *Energy Reports*, 11:736–744, 2024. https://doi.org/10.1016/j.egyr.2023.12.030.

Oliver Molander. Gartner's AI hype cycle is way past its due date — and are we entering a classical ML winter?, 2023. https://olivermolander.medium.com/gartners-ai-hype-cycle-way-passed-its-due-date-and-are-we-entering-a-classical-ml-winter-7c09041c72c4. (accessed 07 April 2025).

National Renewable Energy Laboratory. Wind LCA harmonization, 2013. https://www.nrel.gov/docs/fy13osti/57131.pdf. (accessed 13 April 2025).

Ivan Nikolov, Mikkel Nielsen, Jørgen Garnæs, and Claus Madsen. *Wind Turbine Blade Surfaces Dataset. Mendeley Data, V2* [online], 2020. https;//doi.org/10.17632/jrmm82m4mv.2.

NVIDIA. NVIDIA jetson for next-generation robotics, n.d. https://www.nvidia.com/en-gb/autonomous-machines/embedded-systems/. (accessed 30 March 2025).

OpenAI. Introducing chatGPT, 2022. https://openai.com/index/chatgpt/. (accessed 31 March 2025).

OpenAI, Josh Achiam, Steven Adler, Sandhini Agarwal, Lama Ahmad, Ilge Akkaya, Florencia Leoni Aleman, Diogo Almeida, Janko Altenschmidt, Sam Altman, et al. Gpt-4 technical report. *arXiv preprint arXiv:2303.08774*, 2024.

ORE Catapult. Artificial intelligence – driving forward the future of offshore wind deployment, 2024. https://ore.catapult.org.uk/resource-hub/case-studies/artificial-intelligence-driving-forward-the-future-of-offshore-wind-deployment. (accessed 31 March 2025).

Adam Paszke, Sam Gross, Francisco Massa, Adam Lerer, James Bradbury, Gregory Chanan, Trevor Killeen, Zeming Lin, Natalia Gimelshein, Luca Antiga, et al. Pytorch: An imperative style, high-performance deep learning library, 2019. https://arxiv.org/abs/1912.01703.

Juhi Patel, Lagan Sharma, and Harsh S Dhiman. Wind turbine blade surface damage detection based on aerial imagery and VGG16-RCNN framework. *arXiv preprint arXiv:2108.08636*, 2021.

Tekla S Perry. Move over, Moore's law: Make way for Huang's law, 2018. https://spectrum.ieee.org/move-over-moores-law-make-way-for-huangs-law. (accessed 30 March 2025).

Abdur Rashid, Parag Biswas, Angona Biswas, MD Abdullah Al Nasim, Kishor Datta Gupta, and Roy George. Present and future of AI in renewable energy domain: A comprehensive survey, 2024. https://arxiv.org/abs/2406.16965.

Alexandre Rekeraho, Daniel Cotfas, Petru Cotfas, Titus Balan, Emmanuel Tuyishime, and Rebecca Acheampong. Cybersecurity challenges in IoT-based smart renewable energy. *International Journal of Information Security*, 23:1–17, 2023. https://doi.org/10.1007/s10207-023-00732-9.

Renewable UK. 2024 offshore wind industrial growth plan, 2024. https://www.renewableuk.com/news-and-resources/publications/offshore-wind-industrial-growth-plan/. (accessed 31 March 2025).

Marco Tulio Ribeiro, Sameer Singh, and Carlos Guestrin. "Why should I trust you?": Explaining the predictions of any classifier, 2016. https://arxiv.org/abs/1602.04938.

Callum Rothon, Simon Keizer, Rama Doddipatla, and Nina Dethlefs. Towards interactive anomaly detection using natural language, 2024. https://bda-hull.github.io/publications/Interactive Anomaly Detection using NL IWSDS .pdf.

Olga Russakovsky, Jia Deng, Hao Su, Jonathan Krause, Sanjeev Satheesh, Sean Ma, Zhiheng Huang, Andrej Karpathy, Aditya Khosla, Michael Bernstein, et al. ImageNet large scale visual recognition challenge, 2014. https://arxiv.org/abs/1409.0575.

Ramprasaath R Selvaraju, Michael Cogswell, Abhishek Das, Ramakrishna Vedantam, Devi Parikh, and Dhruv Batra. Grad-Cam: Visual explanations from deep networks via gradient-based localization. *International Journal of Computer Vision*, 128(2):336–359, 2019. https://doi.org/10.1007/s11263-019-01228-7.

ASM Shihavuddin and Xiao Chen. Dtu-drone inspection images of wind turbine. *Mendeley Data, V2,* 2018. https://doi.org/10.17632/hd96prn3nc.2.

Eli Simhayev, Kashif Rasul, and Niels Rogge. Yes, transformers are effective for time series forecasting (+ autoformer), 2023. https://huggingface.co/blog/autoformer.

Jason Staggs, David Ferlemann, and Sujeet Shenoi. Wind farm security: Attack surface, targets, scenarios and mitigation. *International Journal of Critical Infrastructure Protection*, 17:3–14, 2017. https://doi.org/10.1016/j.ijcip.2017.03.001.

The Crown Estate. Offshore wind report 2023, 2024. https://www.thecrownestate.co.uk/our-business/marine/offshore-wind-report-2023. (accessed 31 March 2025).

Hugo Touvron, Thibaut Lavril, Gautier Izacard, Xavier Martinet, Marie-Anne Lachaux, Timothée Lacroix, Baptiste Rozière, Naman Goyal, Eric Hambro, Faisal Azhar, et al. Llama: Open and efficient foundation language models, 2023. https://arxiv.org/abs/2302.13971.

Liam Tung. Google AI on raspberry Pi: Now you get official tensorflow support, 2018. https://www.zdnet.com/article/google-ai-on-raspberry-pi-now-you-get-official-tensorflow-support/

Eamonn Tuton, Xinhui Ma, and Nina Dethlefs. Intelligent digital twin – machine learning system for real-time wind turbine wind speed and power generation forecasting. *E3S Web Conferences*, 433:01008, 2023. https://doi.org/10.1051/e3sconf/202343301008.

Eamonn Tuton, Xinhui Ma, and Nina Dethlefs. Redefining digital twins – a wind energy operations and maintenance perspective. *Journal of Physics: Conference Series*, 2767(5):052001, 2024. https://doi.org/10.1088/1742-6596/2767/5/052001.

Harm van den Brink. Hacking wind turbines—explained, 2022. https://harmvandenbrink.medium.com/hacking-wind-turbines-explained-230997db62f6. (accessed 04 April 2025).

Connor Walker, Callum Rothon, Koorosh Aslansefat, Yiannis Papadopoulos, and Nina Dethlefs. A deep learning framework for wind turbine repair action prediction using alarm sequences and long short term memory algorithms, 2022. https://arxiv.org/abs/2207.09457.

Connor Walker, Callum Rothon, Koorosh Aslansefat, Yiannis Papadopoulos, and Nina Dethlefs. Safety monitoring for large language models: A case study of offshore wind maintenance, 2023. https://hull-repository.worktribe.com/output/4734261.

Connor Walker, Callum Rothon, Koorosh Aslansefat, Yiannis Papadopoulos, and Nina Dethlefs. Using large language models to recommend repair actions for offshore wind maintenance. *Journal of Physics: Conference Series*, 2875(1):012025, 2024a. https://doi.org/10.1088/1742-6596/2875/1/012025.

Connor Walker, Callum Rothon, Koorosh Aslansefat, Yiannis Papadopoulos, and Nina Dethlefs. Safellm: Domain-specific safety monitoring for large language models: A case study of offshore wind maintenance, 2024b. https://arxiv.org/abs/2410.10852.

Anping Wan, Zhipeng Gong, Chao Wei, Khalil AL-Bukhaiti, Yunsong Ji, Shidong Ma, and Fareng Yao. Multistep forecasting method for offshore wind turbine power based on multi-timescale input and improved transformer. *Journal of Marine Science and Engineering*, 12(6):925, 2024. https://doi.org/10.3390/jmse12060925.

Chih-Chiang Wei and Cheng-Shu Chiang. Assessment of offshore wind power potential and wind energy prediction using recurrent neural networks. *Journal of Marine Science and Engineering*, 12(2):283, 2024. https://doi.org/10.3390/jmse12020283.

Simon Willison. Slop is the new name for unwanted AI-generated content, 2024. https://simonwillison.net/2024/May/8/slop/. (accessed 30 March 2025).

Zhenjie Wu, Yulu Zhang, Xiang Wang, Haofei Li, Yufang Sun, and Gang Wang. Algorithm for detecting surface defects in wind turbines based on a lightweight YOLO model. *Scientific Reports*, 14:24558, 2024. https://doi.org/10.1038/s41598-024-74798-3.

Jinyun Yu, Kaipei Liu, Liang Qin, Qiang Li, Feng Zhao, Qiulin Wang, Haofeng Liu, Boqiang Li, Jing Wang, and Kexin Li. DMnet: A new few-shot framework for wind turbine surface defect detection. *Machines*, 10(6):487, 2022.

Yang Yu, Jiahui Wang, Yu Liu, Pingfeng Yu, Dongsheng Wang, Ping Zheng, and Meng Zhang. Revisit the environmental impact of artificial intelligence: The overlooked carbon emission source? *Frontiers of Environmental Science & Engineering*, 18:158, 2024. https://doi.org/10.1007/s11783-024-1918-y.

Ailing Zeng, Muxi Chen, Lei Zhang, and Qiang Xu. Are transformers effective for time series forecasting?, 2022. https://arxiv.org/abs/2205.13504.

Xiangrong Zhu, Yuexiang Xie, Yi Liu, Yaliang Li, and Wei Hu. Knowledge graph-guided retrieval augmented generation, 2025. https://arxiv.org/abs/2502.06864.

Li Zou, Anqi Chen, Xinhua Yang, and Yibo Sun. An improved method of AUD-YOLO for surface damage detection of wind turbine blades. *Scientific Reports*, 15:5833, 2025. https://doi.org/10.1038/s41598-025-89864-7.

12

Incremental Drift-aware Learning in Renewable Energy Systems

Garima Sharma and Jonathan LeSage

Application Engineering Group, MathWorks Inc., Plano, TX, USA

12.1 Introduction

12.1.1 Context of PdM and Renewable Energy System Data

In renewable energy systems, such as wind turbines, solar power plants, or hydropower plants, sensor data is collected to monitor equipment health and ensure efficient operation. The sensor data typically includes readings of voltage, temperature, current, or vibration signals captured from various components, which are usually a time-series data (Hammami et al., 2020; Zhu et al., 2024). The PdM solutions rely on these data coming from sensors to predict potential failures and schedule maintenance activities. However, the sensor data varies due to changes in operational conditions, wear and tear in equipment, dirty solar panels, intermittent clouds on photovoltaics, mandated curtailment of generation, frozen instrumentation giving false signals for wind turbines, grid events triggering grid code compliance, and other environmental factors. Over time, these data distributions drift and can lead to inaccurate predictions when fed into the PdM models. Incremental learning techniques, designed to update models continuously as new data becomes available, offer a promising solution. These techniques are particularly valuable for PdM applications in renewable energy systems, where ever-evolving data streams necessitate adaptive and responsive models.

12.1.2 Challenges in PdM for Renewable Energy Systems

PdM in renewable energy resources is often fraught with challenges stemming from the nature of the sensor data and system dynamics. The main challenge is the concept drift, which is a phenomenon related to the change in the distribution

Renewable Energy Transition with Artificial Intelligence: Challenge-driven Solutions, First Edition. Nina Dethlefs and Joyjit Chatterjee. Published 2026 by John Wiley & Sons, Inc.

of sensor data because of aging equipment, environmental shifts, and operational changes. Another challenge is the velocity and volume of the data produced by the equipment, which demand a scalable and efficient algorithm that can handle the unknown data well for predicting equipment health in PdM applications (Lu et al., 2017). Another challenge is data quality issue, such as additional noise, missing values, and outliers, which further complicates the efficiency of the PdM models (Zhu et al., 2024). Renewable energy systems also encounter data imbalance issues, where failure events are rare compared to normal operations, making it difficult to train models that can accurately detect faults. Addressing these challenges requires advanced methods like incremental learning and drift-aware techniques that can adapt to dynamic data environments while maintaining prediction accuracy.

12.1.3 Incremental Drift-aware Learning for PdM

Incremental drift-aware learning is a combination of two concepts, incremental learning and drift-awareness. Incremental learning is a type of machine learning where models are updated with new knowledge without retraining from scratch, and preserving the old information. Hence, incremental learning makes models adaptive to new data and dynamic environment (Lu et al., 2017; Yu et al., 2020). Drift-aware learning focuses on detecting and adapting to concept drift, ensuring models maintain their inference accuracy even as data distributions evolve. For incremental learning, there are four main applications: incremental classification or regression, incremental principal component analysis (PCA), incremental anomaly detection, and incremental drift-aware learning. These incremental frameworks provide efficient systems to handle streaming data for PdM applications. In this chapter, we will focus mainly on incremental drift-aware learning only. For PdM applications, drift-aware learning is particularly critical, as accurate predictions of equipment health depend on the model's ability to respond to changes in sensor data patterns. Approaches like adaptive thresholds, online ensemble methods, and unsupervised drift detection mechanisms are integral to managing drift effectively. By integrating incremental and drift-aware techniques, renewable energy systems can achieve reliable and efficient predictive maintenance, minimizing downtime and enhancing system resilience.

12.2 Fundamentals of Incremental Learning

12.2.1 Definition and Significance

Incremental learning is also called online learning and is a branch of machine learning, but different from the traditional machine learning approaches. Traditional machine learning algorithms are static in nature, which means they are trained on a whole dataset in one go and then used for inference. While incremental learning involves processing data from an incoming data stream, generally in real time with no or little knowledge about the distribution, size, whether the data is labeled or if there is a change in response variables of the incoming data chunk. Figure 12.1 shows the basic process of incremental learning.

The incremental models are more efficient, flexible, and adaptive. This approach is particularly significant in scenarios involving streaming or nonstationary data, such as predictive maintenance (PdM) for renewable energy systems. Incremental learning ensures that models remain accurate and relevant as data evolves due to factors like equipment wear, environmental changes, or system upgrades. Its ability to balance adaptability and stability makes it ideal for handling real-time data streams, reducing computational costs, and minimizing downtime in industrial applications. Table 12.1 summarizes the fundamental differences, emphasizing the adaptability and efficiency of incremental learning in dynamic environments such as predictive maintenance for renewable energy systems.

For example, let's imagine a smart thermostat designed to automatically set the indoor temperature based on factors like ambient temperature, relative humidity,

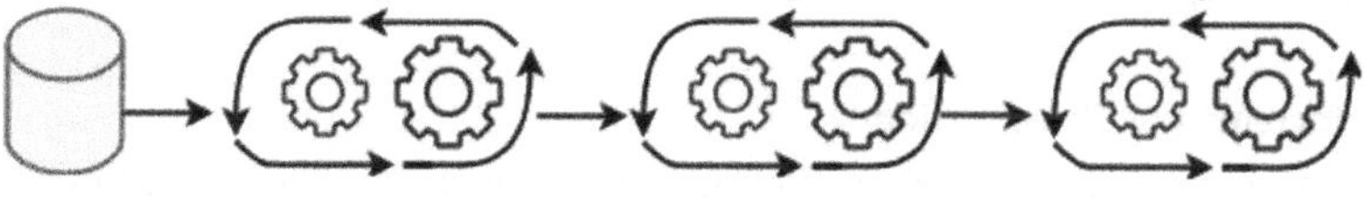

Figure 12.1 Workflow of incremental learning.

Table 12.1 Comparison: incremental learning vs machine learning.

Aspect	Incremental learning	Machine learning
Model training	Continuous on new data chunk	Once on fixed data
Adaptability	Highly adaptable to new data	Low adaptable to new data
Handling drift	Capable of handling	Not capable of handling
Computational cost	Lower	High
Data storage	Adaptive data storage	Fixed data storage
Knowledge retention	Preserves old knowledge	Forgets the old knowledge
Resource efficiency	High	Low
Techniques	Online learning, sliding windows, adaptive ensemble	Batch normalization, ensemble models

time of day, and other environmental inputs. This thermostat also learns the user's specific temperature preferences over time. When the device is first installed, it comes equipped with a preloaded model created by the manufacturer that represents the average person's preferences based on similar measurements.

Once operational, the thermostat collects data continuously, for example, every minute. It uses this information to adjust the temperature according to its initial presets. Over time, it refines its model by learning from the user's interactions—such as manual adjustments or lack of action. This adaptive process can go on indefinitely, allowing the thermostat to become more personalized.

If the thermostat has limited storage for historical data, it must update its model in real time without relying on large amounts of past data. In cases where the thermostat is not preloaded with a model by the manufacturer, it needs to frequently retrain itself from scratch using the data it collects, making real-time learning even more critical.

In this chapter, we will also discuss the solutions provided by MATLAB® software for incremental drift-aware learning. The case study in Section 12.5 will discuss the workflow present in MATLAB software for incremental drift-aware learning used for PdM applications.

12.2.2 Key Approaches in Incremental Learning

There are four main fields where incremental learning could be employed: incremental classification, incremental PCA, incremental anomaly detection, and incremental drift-aware learning.

12.2.2.1 Incremental Classification

In incremental classification, new samples or new classes are added with every chunk of data, and the model is retrained with this new information. Figure 12.2 shows the concept of incremental classification. In MATLAB software, there are several machine learning models present for incremental classification and incremental regression tasks including linear and kernel-based machine learning models. A few of the machine learning models for incremental classification present in MATLAB software are `incrementalClassificationLinear`, `incrementalClassificationKernel`, `incrementalClassificationECOC`, and more. These functions are part of the Statistics and Machine Learning Toolbox™ of MATLAB software.

12.2.2.2 Incremental PCA

In feature engineering, PCA is an important technique used in feature reduction. Incremental principal component analysis (PCA) is an extension of traditional PCA that allows for real-time, efficient dimensionality reduction on streaming or large-scale data. Unlike standard PCA, which processes the entire dataset at once,

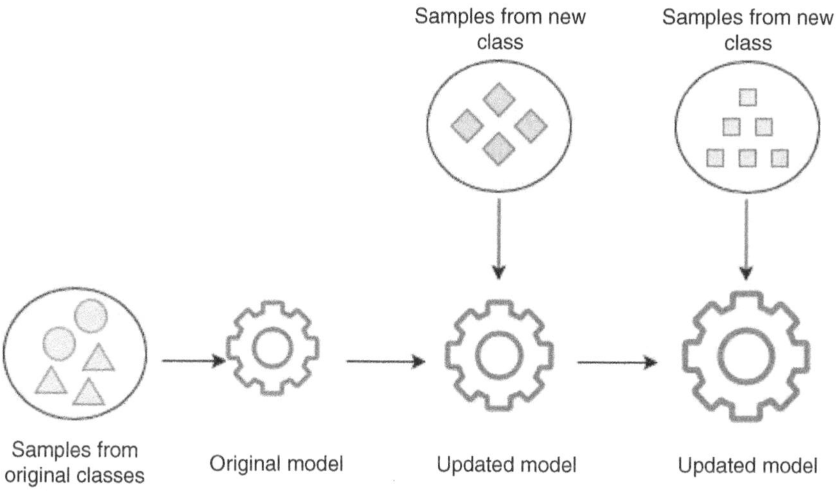

Figure 12.2 Incremental models getting trained on new classes while preserving the older classes.

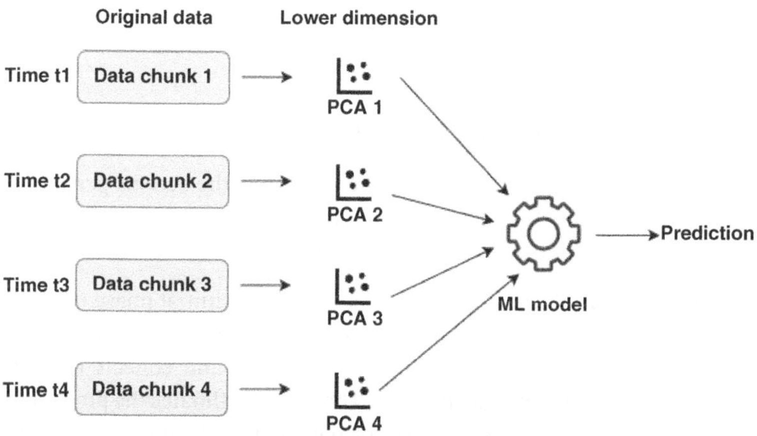

Figure 12.3 Incremental PCA reduces data chunks from high dimension to low dimension, and the model is trained on low-dimension data incrementally.

incremental PCA updates principal components iteratively as new data becomes available, without requiring access to previously seen data. This makes it ideal for scenarios with limited memory or dynamic data, such as predictive maintenance and anomaly detection. Incremental PCA uses batch updates or online techniques to adjust the decomposition, enabling continuous adaptation to evolving data distributions while preserving computational efficiency and scalability. Figure 12.3

shows the concept of incremental PCA. In MATLAB software, the supported function is `incrementalPCA`.

12.2.2.3 Incremental Anomaly Detection

These methods learn from incoming data in small batches or online, allowing the model to detect emerging anomalies while retaining knowledge of past trends. Incremental anomaly detection is particularly effective in addressing concept drift, where normal behavior evolves over time. It balances adaptability with stability, ensuring accurate fault detection in applications such as sensor monitoring and industrial system diagnostics. Incremental anomaly detection is a two-step workflow in MATLAB software, expressed as:

1. **Detect Anomaly:** Use the MATLAB® software's function `isanomaly` to compare the scores of the incoming observations to the set threshold and check if an anomaly is present or not.
2. **Retrain the model:** Update the model by training it on the incoming observations, computing scores, and updating the score threshold.

The machine learning models available in MATLAB software that are suitable for incremental anomaly detection are `incrementalOneClassSVM` and incrementalRobustRandomCutForest.

12.2.3 Incremental Learning Periods in MATLAB

In MATLAB software, when chunks of data are incoming, the actions performed by the incremental learning functions depend on the current state of the model. There are mainly three actions performed by an incremental model, and they are:

1. **Estimation period:** The estimation period marks the initial phase of incremental learning, where the model begins to learn from incoming data. During this phase, the model initializes parameters and collects baseline statistics needed for incremental updates. The model estimates the predictor's moments, such as mean and variance for data standardization, adjusts the learning rates for the solvers, and more. This period is critical for setting the foundation of the model's structure and readiness for real-time adaptation. While not optimized for immediate predictions, the estimation period ensures the model is equipped to process new data effectively, laying the groundwork for robust incremental learning in subsequent phases.
2. **Metrics warm-up:** The metrics warm-up period follows the estimation phase and focuses on fitting the model on the incoming data. After fitting the model on the new data, the metrics such as coefficients of the linear model and data distribution are updated (Pai et al., 2024).

3. **Evaluation period:** The evaluation period represents the mature phase of incremental learning, where the model is assessed on its ability to generalize and maintain performance over time. During this phase, the model handles streaming data, adapts to concept drift, and incorporates new information while retaining prior knowledge. The performance metrics, such as entropy loss or mean square errors, are calculated for each chunk of the data and cumulatively for all the data received. In this phase, the tracking of these metrics ensures the accuracy and reliability of the model in this dynamic environment. This period is where the incremental learning system operates in real-world conditions, providing actionable insights or predictions. Continuous evaluation during this phase ensures that the model adapts effectively to changing data distributions.

12.2.4 Challenges in Incremental Learning

Incremental learning offers a flexible approach to model adaptation, but it also comes with a set of challenges. The two main challenges are catastrophic forgetting and need of adaptive data storage. Data arrives in a stream, and the sample size is unknown and possibly large, which makes data storage difficult. Therefore, the incremental learning algorithm must process the data when they are available and before they get discarded. Hence, there is a need for adaptive data storage that can be increased or decreased as the size of the input data changes. In catastrophic forgetting, incremental models may lose previously learned information when integrating new data. Balancing the retention of old knowledge while learning new patterns is a significant challenge. This problem can be solved by retraining the model on a subset of the previous data so that the model can retain the previous knowledge. Like any other machine learning models, there are some general challenges too, such as noisy data, small data problem, data imbalance, concept drift, hyperparameter tuning, and labeling of the data. Addressing these challenges requires a combination of robust algorithm design, efficient resource management, and dynamic evaluation strategies.

12.3 Concept Drift and Its Effect

12.3.1 What is Concept Drift?

In PdM, concept drift is a phenomenon where statistical parameters of the data coming from the sensors or components change over time because of operating conditions, health of equipment, evolving failure patterns, and environmental factors. Concept drift is particularly challenging as it impacts the reliability of models that monitor and predict equipment health. For instance, a wind turbine may exhibit changing vibration patterns over time due to gradual degradation, requiring models to adapt to these shifts to ensure accurate fault detection. Without addressing concept drift, PdM models can become obsolete, leading to increased false positives, missed failures, and reduced system reliability. There are mainly four types of concept drift:

- **Sudden drift:** This kind of drift occurs abruptly when the data distribution changes instantaneously, such as when a new sensor is added to the system or scheduled maintenance alters equipment behavior.
- **Gradual drift:** Gradual drift happens over time, such as gradual wear and tear of machinery leading to changes in sensor readings. This kind of drift is not easy to track and needs a long time monitoring and tracking of the data.
- **Incremental drift:** In incremental drift, the data distribution changes in small, continuous steps, often reflecting slow degradation in equipment performance.
- **Recurring drift:** This kind of drift happens periodically such as seasonal variations in equipment use or environmental changes.

Each type of drift presents unique challenges for PdM systems and requires tailored strategies to ensure accurate predictions. Incremental learning offers a robust framework for handling concept drift in PdM by enabling models to update dynamically as new data becomes available. MATLAB software's Statistics and Machine Learning Toolbox provides tools specifically designed to manage concept drift in real-time data streams. The main capabilities of this toolbox include:

- **Incremental models:** MATLAB software provides incremental versions of commonly used machine learning models such as linear regression, logistic regression, and decision trees. These models can be updated iteratively as new data arrives, allowing them to adapt to changes in data distributions.
- **Drift detection mechanism:** MATLAB software enables integration of drift detection techniques such as sliding windows, cumulative sum tests, and statistical tests to identify changes in data distributions. These methods can signal when a model update is necessary, ensuring timely adaptation to concept drift.
- **Customizable workflows:** MATLAB software provides a flexible and succinct workflow to update the model. In the next few sections, we will learn about these workflows in detail.
- **Real-time performance metrics:** MATLAB software provides tools for monitoring metrics like accuracy, precision, recall, and error rates in real time. These metrics help identify performance degradation due to concept drift, guiding corrective actions.

12.3.2 Impact of Concept Drift in Renewable Energy

In renewable energy systems, where data is highly dynamic, concept drift could significantly impact the performance of renewable energy systems. Renewable energy systems, such as wind turbines, solar panels, and hydroelectric plants, rely on predictive models for efficient operation, maintenance, and forecasting. Concept drift can lead to a decline in the performance of these models, affecting decision-making and system reliability. Key effects of concept drift in renewable energy include:

- Reduced forecasting accuracy
- Ineffective predictive maintenance (PdM) models
- Energy grid instability
- Economic losses
- Challenges in system optimization

12.4 Drift-aware Learning: Approaches and Techniques

12.4.1 How to Detect Concept Drift and MATLAB Software Solutions

To ensure resilience, renewable energy systems must adopt drift-aware learning techniques that continuously adapt to changing data distributions. Several methods have been developed for drift detection, ranging from statistical tests to machine learning-based approaches. Below is an in-depth exploration of the key techniques, mathematical foundations, and MATLAB software solutions for detecting concept drift.

- **Statistical hypothesis testing:** Statistical tests can detect concept drift by comparing the distributions of historical and recent data samples. Key techniques include:
  - **Kolmogorov–Smirnov (KS) Test:** The KS test is a nonparametric test of equality that can be used to check if a sample belongs to a specific probability distribution or to test if two samples are from the same distribution. Equation 12.1 shows the mathematical formula for KS test.

    $$D = sup_x |F_1(x) - F_2(x)| \tag{12.1}$$

    Here, $F_1(x)$ and $F_2(x)$ are the empirical cumulative distribution functions of two data chunks. If D exceeds a set threshold, a drift is detected. MATLAB software's solution for the chi-square test is employed by the function `kstest2`.
  - **Chi-square test:** Pearson's chi-square test is used to compare observed and expected frequencies in categorical data, especially when the data is large.

Equation 12.2 shows the mathematical equation to find the value of χ^2.

$$\chi^2 = \sum \frac{(O_i - E_i)^2}{E_i} \tag{12.2}$$

Here, O_i and E_i are the observed and expected frequencies, respectively. Based on the value of χ^2 compared to a threshold, the null hypothesis is accepted or rejected. MATLAB software's solution for chi-square test is employed by the function `chi2gof`.

- **Window-based approaches:** Window-based methods analyze data streams in segments, comparing statistics from a reference window (past data) and a test window (recent data). Key window-based techniques to detect drift include:
  - **Sliding window:** The sliding window method for drift detection is used by continuously moving the window and comparing the latest windowed data distribution with the previous one, and detects a potential drift if there is a significant change in the data distribution between two windows. This technique maintains a fixed-size window and replace old data with new data as it arrives. The metric to compare could be a mean, variance, standard deviation, or classification accuracy. MATLAB software enables sliding window analysis using `movmean` and `movstd`.

$$\Delta = |\text{Metric}(W_{\text{reference}}) - \text{Metric}(W_{\text{test}})| \tag{12.3}$$

where a significant Δ reflects a drift in the incoming stream.

  - Adaptive windowing: In this windowing method, the size of the window adjusts based on the changes in the data distribution. It splits the window into two sub-windows, i.e. recent and historical, and compares their means to detect drift. Using Hoeffding's inequality, drift is identified when the difference between means of the two windows exceeds a threshold.

$$|\mu_1 - \mu_2| > \epsilon \tag{12.4}$$

- **Drift detection methods:** The Drift Detection method (DDM) is a widely used method for detecting concept drift in streaming data, particularly for classification tasks. It works by monitoring the error rate of a machine learning model over time. If the error rate increases significantly, it indicates potential drift in the data distribution, which can degrade model performance. In MATLAB software, the functions that can be used to find DDM are `detectdrift` and `incrementalConceptDriftDetector`. The function `incrementalConceptDriftDetector` instantiates an incremental concept drift detector object, function `detectdrift` updates the internal statistics and the drift status with new data and returns the modified detector.

- **Hoeffding Drift Detection Method:** The Hoeffding Drift Detection Method (HDDM) is a statistical approach for concept drift detection based on Hoeffding's inequality, which provides an upper bound on the probability of the mean of a sample deviating significantly from the true mean of the distribution. There are two types of HDDM methods in MATLAB software. One HDDM method is based on moving average (A-test) and called the Hoeffding Drift Detection Method moving average (HDDMA) algorithm. The other is based on exponential weighted moving average (W-test), and hence is called Hoeffding Drift Detection Method exponential weighted moving average algorithm (HDDMW). In MATLAB software, the supporting function is `HoeffdingDriftDetectionMethod`.

  The Eq. below shows Hoeffding's inequality:

  $$P(|\hat{\mu} - \mu| > \epsilon) \leq 2\exp(-2n\epsilon^2) \tag{12.5}$$

  Where, $\hat{\mu}$ is sample mean, μ is true mean, n is sample size, and ϵ is allowed deviation. HDDM uses this bound to detect deviations between the sample mean of the reference window and the test window. If the probability exceeds a predefined threshold, drift is detected.

  While the aim of DDM and HDDM is to detect drift, but there is a significant difference between their methodologies, such as:

  - **Focus on statistical bounds:** HDDM directly leverages Hoeffding's inequality to bound deviations, making it statistically more robust. While DDM focuses on tracking error rates in a classification model, identifying drift when error deviations exceed a threshold.
  - **Detection criteria:** HDDM uses the difference in observed means to detect changes, while DDM evaluates error rates and their variance over time.
  - **Flexibility:** HDDM is model-agnostic and purely data-driven, while DDM is inherently tied to supervised learning models.

12.5 Case Study: Detection of Drift Using MATLAB Software Solutions

In this section, we will explore the MATLAB software capabilities in the field of incremental drift-aware learning by following an example of drift detection in the numerical synthetic data generated by MATLAB. The synthetic data generated has 40 predictors, 20 000 observations, and a logical response variable, where "0" represents "No failure" and "1" represents "Failure" reported in the component. The MATLAB software codes and their respective outputs are discussed with each section of the code below. The MATLAB software code written below was created on `R2024b` version of MATLAB software.

Let's generate the synthetic data and analyze its size. To prepare the data for drift detection, save the predictors and response variables in different variables. The predictors are saved in the variable `XTrain` and responses are saved in the variable `YTrain`.

```
% Generate synthetic random and normalized data
numRows = 20000;
numCols = 40;
synthetic_data = randn(numRows,numCols);
% Generate random logical array representing predictors
classLables = randi([0,1], numRows,1);
XTrain = synthetic_data;
YTrain = classLabels;
% display number of observations and variables
fprintf('The number of observations are %d and number of variables
 are %d.', size(XTrain,1), size(XTrain,2))
```

Output:

`The number of observations is 20 000 and the number of variables is 40.`

In the next step, allocate the number of samples in each data chunk and the number of iterations to estimate the presence of drift in the chunks of streaming data. The number of observations per data chunk is chosen to be say 50 in this example. The MATLABcode is:

```
n = height(XTrain);
% select number of obervations in each data chunk
numObsPerChunk = 50;
% calculate number of chunks
nchunk = floor(n/numObsPerChunk);
fprintf("The number of data chunks are %d and each chunk has %d
 number of observations", nchunk, numObsPerChunk)
```

Output:

The number of data chunks is 400, and each chunk has 50 observations.

Let's create a baseline incremental Naive Bayes model for binary classification, an incremental concept drift detector that uses the Hoeffding's Bounds Drift Detection Method with moving average (HDDMA) and is able to detect the drift in the data, and an incremental drift-aware model that works with the base learner Naive Bayes model and drift detector model. Specify the training period as 1000 observations. The model will be trained for first 1000 samples and will not perform any classification for this period. Also, specify additional name-value pairs such as stable count limit as 1000 and warning count limit as 1400.

```
% base learner model with 2 classes
incMdl = incrementalClassificationNaiveBayes(ClassNames=[0 1])
% incremental drift aware learner with training period of 1000
 samples
idaMdl = incrementalDriftAwareLearner(incMdl,TrainingPeriod=1000,
 StableCountLimit=1000,WarningCountLimit=1400)
```

```
incMdl =
  incrementalClassificationNaiveBayes

                      IsWarm: 0
                     Metrics: [1×2 table]
                  ClassNames: [0 1]
              ScoreTransform: 'none'
           DistributionNames: 'normal'
      DistributionParameters: {2×0 cell}


  Properties, Methods
```

Figure 12.4 Details of incremental Naive Bayes classifier, which is a base learner for incremental drift-aware learner model.

```
idaMdl =
  incrementalDriftAwareLearner

           IsWarm: 0
          Metrics: [1×2 table]
      BaseLearner: [1×1 incrementalClassificationNaiveBayes]
    DriftDetector: [1×1 HoeffdingDriftDetectionMethod]
       IsTraining: 1


  Properties, Methods
```

Figure 12.5 Incremental drift-aware detection model.

Output: Figures 12.4 and 12.5 show the details of the incremental Naive Bayes classifier and incremental drift-aware learner.

To track the performance of the incremental drift-aware learner, allocate the variables `resultsTable` to save the cumulative and window-based classification error for each data chunk.

```
% Create table to store results
resultsTable = array2table(zeros(nchunk,2),'VariableNames',['
Cumulative' 'Window']);
resultsTable.Status = zeros(nchunk,1);
resultsTable.StatusName = strings(nchunk,1);
```

Output:
A variable `resultsTable` to capture the status of the drift-aware models (0,1, or 2), its property `StatusName` captures the state of the drift as one of the categories: stable, warning, or drift for cumulative and window-based.

Simulate a data stream with incoming chunks of 50 observations each, and perform incremental drift-aware learning. At each iteration, the following steps are performed:

- Call `updateMetricsAndFit` to update the performance metrics and fit the incremental drift-aware model to the incoming data.
- Track and record the drift status and the classification error for visualization purposes.

```
% Loop through chunks (simulating streaming), update metrics and
fit incremental learner
for j = 1:nchunk
    % Set index range for this chunk
    ibegin = min(n,numObsPerChunk*(j-1)+1);
    iend   = min(n,numObsPerChunk*j);
    idx = ibegin:iend;

    % Update the drift-aware model
    idaMdl = updateMetricsAndFit(idaMdl,XTrain(idx,:),YTrain(idx));

    % Record results in the jth row of the table
    metrics = idaMdl.Metrics{"MinimalCost",:};
    statusName = string(idaMdl.DriftStatus);
    if idaMdl.DriftDetected
        status = 2;
    elseif idaMdl.WarningDetected
        status = 1;
    else
        status = 0;
    end
    resultsTable(j,:) = table(metrics(1), metrics(2), status,
statusName);
end
% print the last 8 values of resultTable
tail(resultsTable)
```

Output: Figure 12.6 shows the classification errors for the last eight chunks of the data in the stream for both cumulative and window-level.

Cumulative	Window	Status	StatusName
0.49616	0.55292	0	"Stable"
0.49606	0.55292	0	"Stable"
0.49615	0.55292	0	"Stable"
0.49598	0.44572	0	"Stable"
0.49611	0.44572	0	"Stable"
0.49609	0.44572	0	"Stable"
0.49626	0.44572	0	"Stable"
0.49626	0.52264	0	"Stable"

Figure 12.6 Classification error table for both cumulative and window-level.

The `updateMetricsAndFit` function first evaluates the performance of the model by calling `updateMetrics` on incoming data, and then fits the model to data by calling fit: The `updateMetrics` function evaluates the performance of the model as it processes incoming observations. The function writes specified metrics, measured cumulatively and within a specified window of processed observations, to the metrics model property. The fit function fits the model by updating the base learner and monitoring for drift given an incoming batch of data. When you call fit, the software performs the following procedure:

- Trains the model up to `NumTrainingObservations` observations.
- After training, the software starts tracking the model loss to see if any concept drift has occurred and updates drift status accordingly.
- When the drift status is Warning, the software trains a temporary model to replace the `BaseLearner` in preparation for an imminent drift.
- When the drift status is Drift, the temporary model replaces the `BaseLearner`.
- When the drift status is Stable, the software discards the temporary model.

Now, plot the cumulative and per-window classification error for all 400 iterations. Mark the warmup and training periods on the plot as well.

```
% Plot the results
h = plot( [resultsTable.Cumulative, resultsTable.Window] );
xlim([0 nchunk])
ylabel("Minimal Cost")
xlabel("Iteration")
```

```
% add vertical line markers
xline(idaMdl.MetricsWarmupPeriod/numObsPerChunk,"g-.","Warmup
 Period",LineWidth= 1.5)
xline(idaMdl.TrainingPeriod/numObsPerChunk,"b-.","Training Period",
 ...
LabelVerticalAlignment="middle",LineWidth= 1.5)
legend(h, ["Cumulative", "Window"])
legend(h,Location="best")
title("Incremental Learner with No Drift Detected")
```

Output: Figure 12.7 shows the visualization for all 400 iterations of the streaming data.

Plot the drift status versus the iteration number.

```
figure()
gscatter(1:nchunk,resultsTable.Status,resultsTable.StatusName,'gmr'
 ,'*ox',[4 5 5], ...
'on',"Iteration","Drift Status","filled")
xlim([0 nchunk])
ylim([-1 3])
title("Drift Status")
```

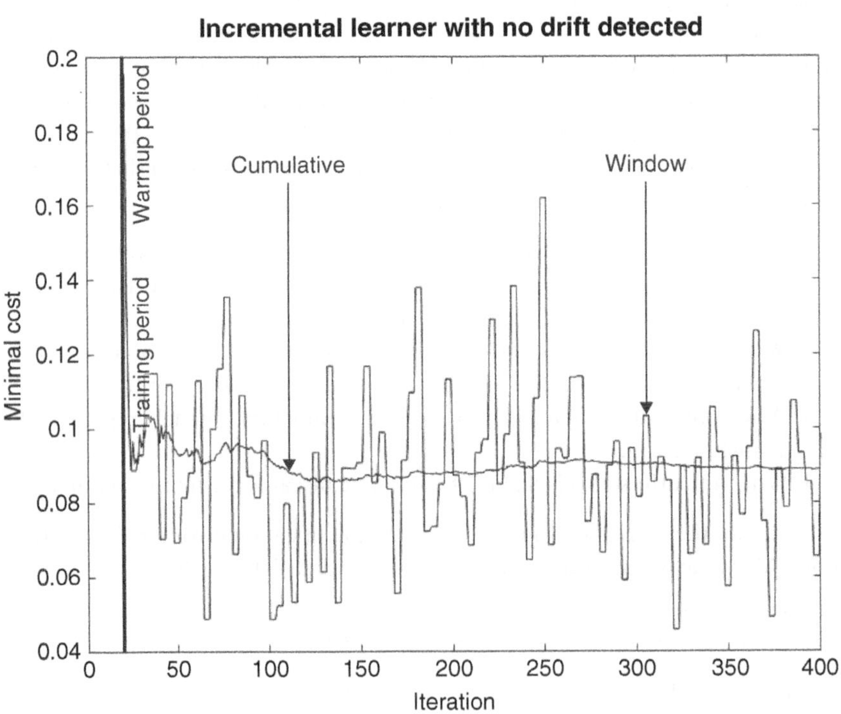

Figure 12.7 Classification error for windowed and cumulative data for no drift status.

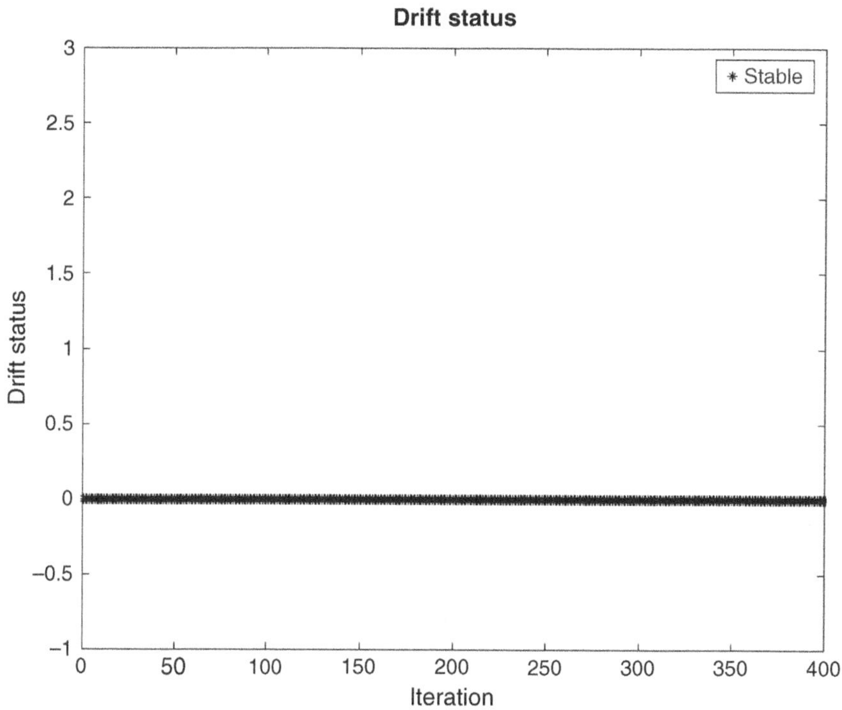

Figure 12.8 No drift detected in the original data.

Output: Figure 12.8 shows that no drift has been predicted by the drift detector as we work on incoming data chunks.

Now, to evaluate the efficacy of the drift detection model we created above, a drift is added at the midpoint of the data. To add the drift, the second half of the predictors is updated as shown below in the MATLABCode:

```
% Modify the second half of X data using X = 10.05*X + 0.5
modifyStart = floor(numel(YTrain)/2);
modifyEnd = numel(YTrain);
XTrain(modifyStart:modifyEnd,:) = XTrain(modifyStart:modifyEnd,:)
*10.05 + 0.5;
n = height(XTrain);
numObsPerChunk = 50;
nchunk = floor(n/numObsPerChunk);
% Reset the model
incMdl = incrementalClassificationNaiveBayes(ClassNames=[0 1]);
idaMdl = incrementalDriftAwareLearner(incMdl,TrainingPeriod=1000,
StableCountLimit=1000,WarningCountLimit=1400);

% Create table to store results
resultsTable = array2table(zeros(nchunk,2),'VariableNames',["
Cumulative" "Window"]);
```

```
resultsTable.Status = zeros(nchunk,1);
resultsTable.StatusName = strings(nchunk,1);

% Loop through chunks (simulating streaming), update metrics and
 fit incremental learner
for j = 1:nchunk
    % Set index range for this chunk
    ibegin = min(n,numObsPerChunk*(j-1)+1);
    iend   = min(n,numObsPerChunk*j);
    idx = ibegin:iend;

    % Update the drift-aware model
    idaMdl = updateMetricsAndFit(idaMdl,XTrain(idx,:),YTrain(idx));

    % Record results in the jth row of the table
    metrics = idaMdl.Metrics{'MinimalCost',:};
    statusName = string(idaMdl.DriftStatus);
    if idaMdl.DriftDetected
        status = 2;
    elseif idaMdl.WarningDetected
        status = 1;
    else
        status = 0;
    end
    resultsTable(j,:) = table(metrics(1), metrics(2), status,
 statusName);
end
```

```
% Plot the results
h = plot( [resultsTable.Cumulative, resultsTable.Window] );

xlim([0 nchunk])
ylabel("Minimal Cost")
xlabel("Iteration")

xline(idaMdl.MetricsWarmupPeriod/numObsPerChunk,"g-.","Warmup
 Period",LineWidth= 1.5)
xline(idaMdl.TrainingPeriod/numObsPerChunk,"b-.","Training Period",
 ...
LabelVerticalAlignment="middle",LineWidth= 1.5)
xline(modifyStart/numObsPerChunk,"m--","X Modified",
 LabelVerticalAlignment="middle", ...
LineWidth= 1.5)
xline((modifyStart+idaMdl.TrainingPeriod)/numObsPerChunk,"b-.","
 Training Period", ...
LabelVerticalAlignment="middle",LineWidth= 1.5)

legend(h,["Cumulative", "Window"])
legend(h,Location="best")

title("Incremental Learner with Drift Detected when X is Modified")
```

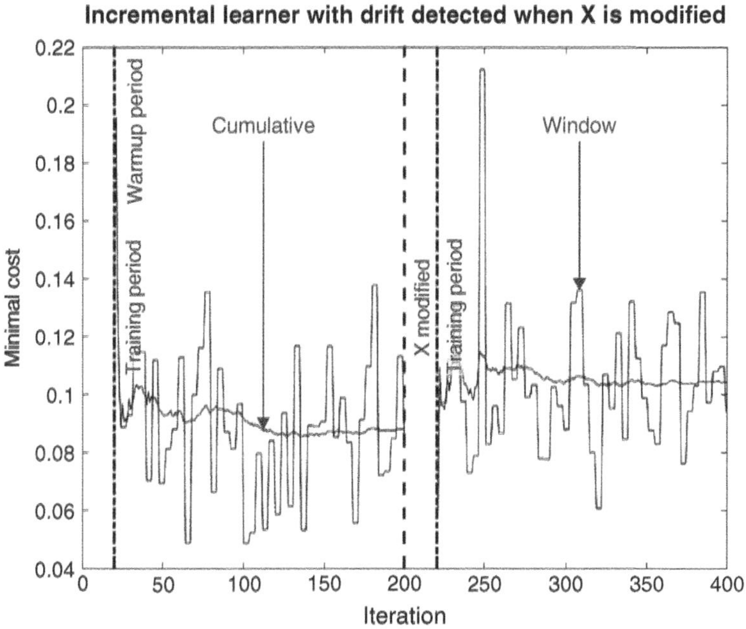

Figure 12.9 Classification error for windowed and cumulative data when drift is added.

Output: Figure 12.9 shows the classification errors observed when a drift is added.

```
figure()
gscatter(1:nchunk,resultsTable.Status,resultsTable.StatusName,'gmr',
'*ox',[4 5 5], ...
'on',"Iteration","Drift Status","filled")
xlim([0 nchunk])
ylim([-1 3])
title("Drift Status")
```

Output: Figure 12.10 shows the detection of drift after the midpoint of the data chunks.

Now for the third case, let's add multiple drifts into the data. To achieve that, concatenate the "XTrain" and "YTrain" data with itself. Now, the number of observations has been doubled, and the data is drifting three times now.

```
% case 3: Add multiple drifts in the data
XTrain = [XTrain;XTrain];
YTrain = [YTrain;YTrain];
fprintf("The number of observations are %d and number of variables
are %d", size(XTrain,1), size(XTrain,2))
```

Output:

```
The number of observations is 40000 and number of
variables is 40.
```

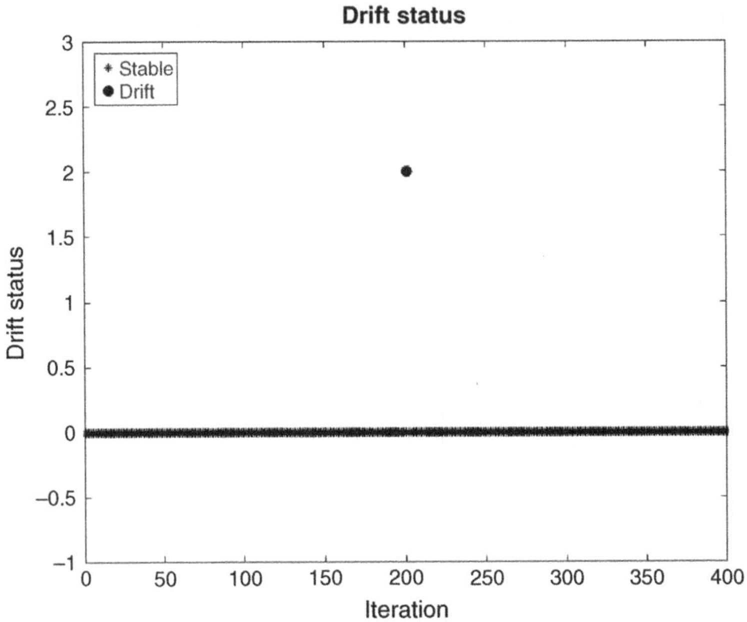

Figure 12.10 A drift is detected at the mid of the data.

Now, pre-allocate the variable `resultsTable` as now the number of observations and hence the number of chunks became double than the original data.

```
n = height(XTrain);
numObsPerChunk = 50;
nchunk = floor(n/numObsPerChunk);
% Create table to store results
resultsTable = array2table(zeros(nchunk,2),'VariableNames',["
Cumulative" "Window"]);
resultsTable.Status = zeros(nchunk,1);
resultsTable.StatusName = strings(nchunk,1);
```

Initialize the base learner model and drift-aware learner models. Then update the metrics for each data chunk by detecting any drifts observed by the detector model.

```
% Reset the model
incMdl = incrementalClassificationNaiveBayes(ClassNames=[0 1]);
idaMdl = incrementalDriftAwareLearner(incMdl,TrainingPeriod=1000,
StableCountLimit=1000,WarningCountLimit=1400);
% Loop through chunks (simulating streaming), update metrics and
fit incremental learner
for j = 1:nchunk
    % Set index range for this chunk
    ibegin = min(n,numObsPerChunk*(j-1)+1);
    iend   = min(n,numObsPerChunk*j);
    idx = ibegin:iend;
```

```
        % Update the drift-aware model
        idaMdl = updateMetricsAndFit(idaMdl,XTrain(idx,:),YTrain(idx));

        % Record results in the jth row of the table
        metrics = idaMdl.Metrics{"MinimalCost",:};
        statusName = string(idaMdl.DriftStatus);
        if idaMdl.DriftDetected
            status = 2;
        elseif idaMdl.WarningDetected
            status = 1;
        else
            status = 0;
        end
        resultsTable(j,:) = table(metrics(1), metrics(2), status,
    statusName);
    end
    %print results for last 8 data chunks
    tail(resultsTable)
```

Output: Figure 12.11 shows the classification error for the last eight chunks of data when multiple drifts are added.

Plot the classification errors for each data chunk and scatter plot to show the drift status recorded.

```
    % Plot the results
    h = plot( [resultsTable.Cumulative, resultsTable.Window] );

    xlim([0 nchunk])
    ylabel("Minimal Cost")
    xlabel("Iteration")

    xline(idaMdl.MetricsWarmupPeriod/numObsPerChunk,"g-.","Warmup
    Period",LineWidth= 1.5)
    xline(idaMdl.TrainingPeriod/numObsPerChunk,"b-.","Training Period",
     ...
    LabelVerticalAlignment="middle",LineWidth= 1.5)
    xline(modifyStart/numObsPerChunk,"m--","X Modified",
    LabelVerticalAlignment="middle", ...
    LineWidth= 1.5)
    xline(2*modifyStart/numObsPerChunk,"m--","X Modified",
    LabelVerticalAlignment="middle", ...
    LineWidth= 1.5)
    xline(3*modifyStart/numObsPerChunk,"m--","X Modified",
    LabelVerticalAlignment="middle", ...
    LineWidth= 1.5)

    legend(h,["Cumulative", "Window"])
    legend(h,Location="best")

    title("Incremental Learner with Drift Detected Multiple Times")
```

Output: Figure 12.12 shows the drifts in the classification errors every time the data is modified.

Cumulative	Window	Status	StatusName
0.08902	0.085613	0	"Stable"
0.089036	0.085613	0	"Stable"
0.088902	0.085613	0	"Stable"
0.08872	0.065586	0	"Stable"
0.088815	0.065586	0	"Stable"
0.088925	0.065586	0	"Stable"
0.088951	0.065586	0	"Stable"
0.088821	0.098333	0	"Stable"

Figure 12.11 Classification error table for both cumulative and window-level when multiple drifts are detected.

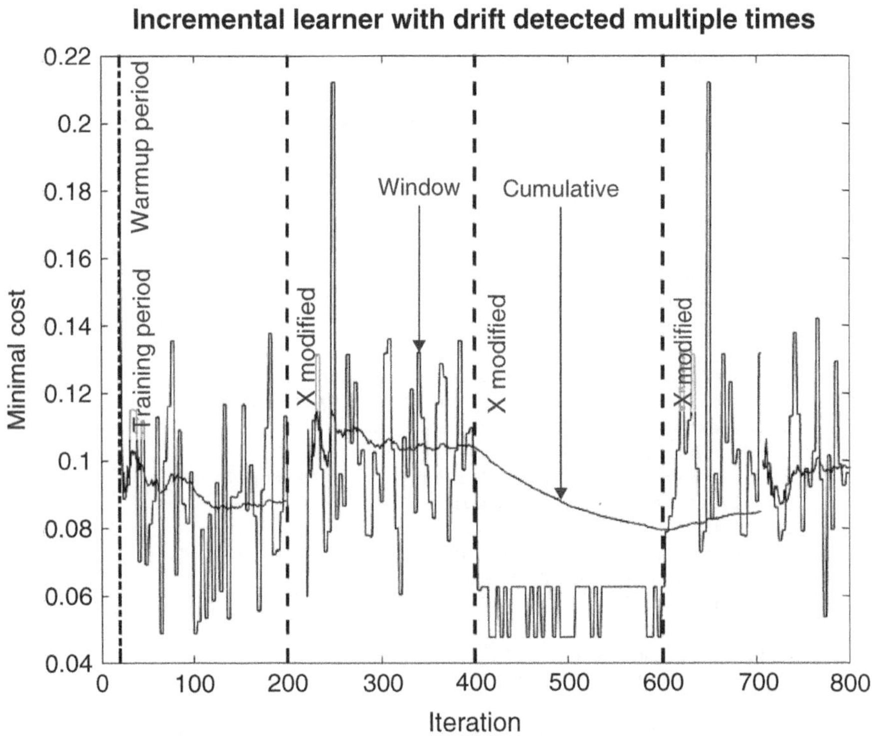

Figure 12.12 Classification error with respect to data chunks reflecting multiple drifts.

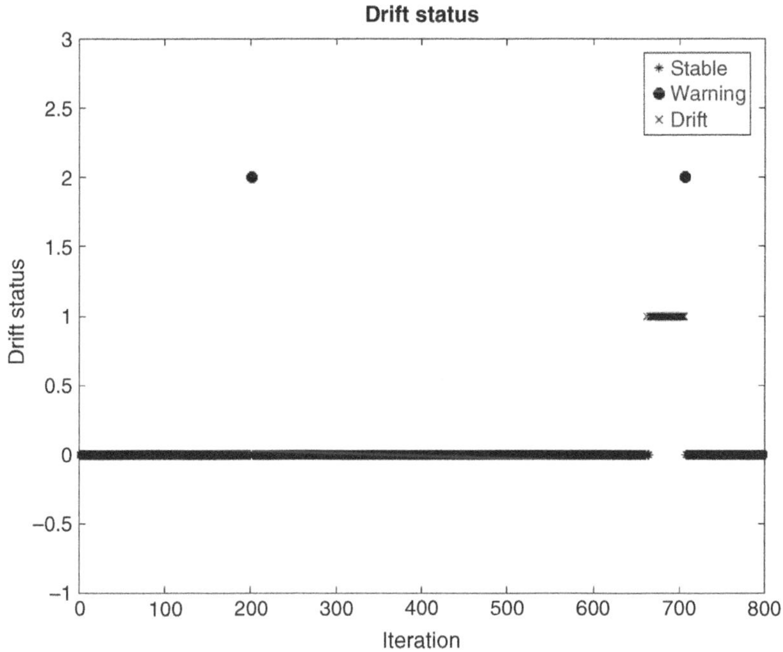

Figure 12.13 Multiple drifts and warnings detected by the drift-aware learner model.

```
figure()
gscatter(1:nchunk,resultsTable.Status,resultsTable.StatusName,'gmr'
,'*ox',[4 5 5], ...
'on',"Iteration","Drift Status","filled")
xlim([0 nchunk])
ylim([-1 3])
title("Drift Status")
```

Output: Figure 12.13 shows the scatter plot reflecting multiple warnings and drifts in the data.

12.6 Conclusion

In renewable energy systems, concept drift is a critical challenge, particularly in applications like PdM. Detecting and adapting to drift ensures the sustained accuracy and reliability of machine learning models. Techniques such as the Drift Detection Algorithm (DDA), Hoeffding Drift Detection Method (HDDM), and Adaptive Windowing offer robust solutions for identifying changes in data distributions. By leveraging statistical properties like error rates, moving averages, and confidence intervals, these methods enable timely detection and adaptation to drift.

MATLAB software's extensive toolbox supports the implementation of these techniques through its data processing, visualization, and live-streaming capabilities. Tools like `detectdrift`, `incrementalConceptDriftDetector`, `movmean`, `movstd`, and more allow seamless integration of drift detection mechanisms. By employing these methods, organizations can build resilient systems capable of handling evolving data streams, ultimately improving decision-making, operational efficiency, and system reliability in real-world applications.

Acknowledgment

We would like to extend our sincere gratitude to The MathWorks Inc. for their invaluable support and the advanced capabilities provided by MATLAB software and the Statistics and Machine Learning Toolbox. These tools greatly facilitated the exploration and implementation of drift detection and incremental learning techniques, enabling the development of this book chapter. The content of the book chapter is owned by the authors, not by The MathWorks Inc.

Bibliography

Zeineb Hammami, Moamar Sayed-Mouchaweh, Wiem Mouelhi, and Lamjed Ben Said. Neural networks for online learning of non-stationary data streams: A review and application for smart grids flexibility improvement. *Artificial Intelligence Review*, 53(8):6111–6154, 2020.

Yang Lu, Yiu Ming Cheung, and Yuan Yan Tang. Dynamic weighted majority for incremental learning of imbalanced data streams with concept drift. In *Proceedings of the 26th International Joint Conference on Artificial Intelligence, IJCAI 2017*, pp. 2393–2399. International Joint Conferences on Artificial Intelligence, 2017.

Yu-Tung Pai, Nien-En Sun, Cheng-Te Li, and Shou-de Lin. Incremental data drifting: Evaluation metrics, data generation, and approach comparison. *ACM Transactions on Intelligent Systems and Technology*, 15, 1–26, 2024.

Lu Yu, Bartłomiej Twardowski, Xialei Liu, Luis Herranz, Kai Wang, Yongmei Cheng, Shangling Jui, and Joost van de Weijer. Semantic drift compensation for class-incremental learning. In *Proceedings of the IEEE/CVF Conference on Computer Vision and Pattern Recognition, Energies, MDPI*, pp. 6982–6991, 2020.

Xinchun Zhu, Yang Wu, Xu Zhao, Yunchen Yang, Shuangquan Liu, Luyi Shi, and Yelong Wu. Overview of wind and photovoltaic data stream classification and data drift issues. *Energies*, 17(17):4371, 2024.

Index

Note: Page numbers in *italics* and **bold** refers to figures and tables respectively.

Printed and bound by CPI Group (UK) Ltd, Croydon, CR0 4YY

17/09/2026

14964232-0003